Placer Gold Deposits of Nevada

By MAUREEN G. JOHNSON

GEOLOGICAL SURVEY BULLETIN 1356

A catalog of location, geology, and production with lists of annotated references pertaining to the placer districts

UNITED STATES GOVERNMENT PRINTING OFFICE, WASHINGTON : 1973

UNITED STATES DEPARTMENT OF THE INTERIOR

ROGERS C. B. MORTON, *Secretary*

GEOLOGICAL SURVEY

V. E. McKelvey, *Director*

Library of Congress catalog-card No. 72–600324.

CONTENTS

ILLUSTRATIONS

TABLES

PLACER GOLD DEPOSITS OF NEVADA

By Maureen G. Johnson

ABSTRACT

One hundred and fifteen placer districts in Nevada are estimated to have produced a minimum of 1,700,000 ounces of placer gold from 1849 to 1968. The location, areal extent, past production, mining history, and probable lode source for each district are summarized on the basis of information obtained from a wide variety of published reports relating to placer deposits. Annotated references to all reports give information about individual deposits for each district.

Most of the placer gold found in Nevada has been derived from veins and replacement deposits that have been successfully worked for the gold and silver content of the ores. In the few districts for which the source of the gold is unknown, it is presumed to be small scattered veins in the adjacent bedrock. In most of the very productive lode mining districts, only small amounts of placer gold have been recovered, whereas in the very productive placer districts, lode-gold production is close to, and sometimes less than, placer gold production. Most of the placer mining was done before 1900 by small-scale methods utilizing portable equipment, such as rockers, sluices, and drywashers, to work small deposits; most of the placer gold produced in the State since 1900 was mined by a few very large dredge operations between 1920 and 1959.

INTRODUCTION

HISTORY OF PLACER MINING IN NEVADA

The first authenticated discovery of placer gold in Nevada was made in 1849 by Abner Blackburn, a member of an emigrant train to California, at the junction of Gold Canyon and the Carson River at the present site of Dayton, Lyon County (De Quille, 1891; Vanderburg, 1936a). Parties of men worked the gravels in Gold Canyon and nearby Six Mile Canyon, Storey County, for 8 years before the source of the placers, the Ophir silver lode, was discovered by Peter O'Reiley and Patrick McLaughlin in 1857 while digging a small water hole for placer mining in Six Mile Canyon (De Quille, 1891). Other lode discoveries in the immediate area followed, and soon the whole world knew of the Comstock lode in Nevada. Although placer mining continued on a small scale in Gold Canyon and Six Mile Canyon, and other placers were discovered elsewhere in the State, the richness and fame of the Comstock

1

lode far overshadowed the importance of placer production and new placer discoveries.

Following the discovery of placers at Gold Canyon, placer discoveries in Nevada were broadly in three periods: the 1860's to 1880's, when many small deposits throughout the State were discovered and sporadically worked and several large placers were discovered and extensively worked; the short period between 1906 and 1910, when very rich placers were discovered at Lynn, Battle Mountain, Manhattan, and Round Mountain; the early 1930's, when economic conditions created by the depression caused a renewed interest in placer mining, and many individuals sought, and a few discovered, new placer areas throughout the State. The location of the placers described in this report is shown on plate 1.

Very little factual information can be found about the early periods of placer mining in Nevada. For many placers, the only reports available are hearsay estimates of production and speculations about the extent of the placer ground based on remnants of placer pits, shafts, and other workings. Many of the placers said to have had a high production between 1860 and 1890 were worked by Chinese miners who came to Nevada during the building of the railroads and stayed on to work at mining and other activities. The Chinese were reputed to be secretive with their earnings from the placers and did not ship the gold to the mint by Wells Fargo or other shippers. They worked the gravels very thoroughly in areas where American miners did not wish to expend great labor to win the gold. The placers in the Sierra and Spring Valley districts, Pershing County, were worked by Chinese miners; they have a very high estimated production before 1900 and a comparatively low known production since that time.

One reason for the lack of information about early placer-mining activity in Nevada was the great attention given to the rich silver-lode districts such as the Comstock, Eureka, and Reese River districts. Whereas in many other States, the discovery of gold placers stimulated the search for lode-gold deposits and other gold placers, in Nevada early attention was devoted to searching for rich silver lodes not necessarily associated with derived placers.

The comparatively late discovery of some of the richest placers in the State has afforded a very clear picture of the development of placer mining during the 1900's. The discovery of rich silver ores at Tonopah in 1900 and rich gold ores at Goldfield in 1902 stimulated great activity in mining exploration throughout Nevada. Many placers discovered during the 1906–10 period were found by men looking for ores similar to ores at Tonopah and Goldfield. Placer mining at Manhattan and Round Mountain districts, Nye County, and Battle Mountain district, Lander County, began with numerous small drywash operations in the gravels, then expanded as water supplies were developed for sluicing and hydraulic methods of mining. Late in the history of these districts, but long after many other placer districts were inactive, large-scale dredging operations began. The success of the dredge operations in these semiarid districts is unique in the history of placer mining in the Southwestern States.

Placer-mining history in the other districts is typical of desert placer mining throughout the southwest. Most production resulted from the relatively intense period of prospecting immediately following discovery; a decline in placer-mining activity followed, then a small revival during the early 1930's. The economic depression of the early 1930's stimulated investigations of many Nevada placer districts for the purpose of developing large-scale placer-mining operations. By the late 1930's, many mining companies had investigated many placer areas and had formulated plans to develop certain areas. The placer activity of the 1930's was abruptly halted by the beginning of World War II and the passage of War Board Order L–208, which restricted gold mining throughout the country. The dredge operation at Manhattan was given special permission to continue operations, although on a reduced scale, and, as a result, placer gold production after 1942 did not decline as markedly in Nevada as in other States.

Most of the placer mining was done by the basic methods of dry-washing, sluicing, and rarely, small-scale hydraulic mining. In addition to the large dredge operations at Manhattan, Round Mountain, and Battle Mountain, other dredges operated in different districts, notably Silver City, Lyon County; Spring Valley, Pershing County; Van Duzer, Elko County; and Bullion, Lander County. Since the completion of the Round Mountain dredge operation in 1959, placer mining in Nevada has progressively diminished in importance.

PURPOSE AND SCOPE OF PRESENT STUDY

The present paper is a compilation of published information relating to the placer gold deposits of Nevada, one of a series of four papers describing the gold placer deposits in the Southwestern States. The purpose of the paper is to outline areas of placer deposits in the State and to serve as a guide to their location, extent, production history, and source. This work was undertaken as part of the investigation of the distribution of known gold occurrences in the Western United States.

Each placer is described briefly. Location is given by geographic area and township and range. Topographic maps and geologic maps which show the placer area are listed for each. Access to each area is indicated by direction and distance along major roads and highways from a nearby center of population.

Detailed information relating to the exact location of placer deposits, their thickness, distribution, and average gold content (all values cited in the text have been converted to gold at $35 per ounce, except where otherwise noted) is included, when available, under "Extent." U.S. Bureau of Land Management land plats of surveyed township and ranges were especially helpful in locating old placer claims and some creeks and dry-washes not named on recent topographic maps. These land plats were consulted for all the surveyed areas in Nevada. U.S. Bureau of Mines records were also consulted for the location of small placer claims.

Discovery of placer gold and subsequent placer-mining activity are briefly described under "Production history." Detailed discussion of min-ing operations is omitted, as this information can be found in the indi-vidual papers published by the State of Nevada, in the yearly Mineral Resources and the Minerals Yearbooks published by the U.S. Bureau of Mines and the U.S. Geological Survey, and in many mining journals. Placer gold production, in ounces (table 1), was compiled from the yearly Mineral Resources and Minerals Yearbook volumes and from infor-mation supplied by the U.S. Bureau of Mines. These totals of recorded production are probably lower than actual gold production, for substan-tial amounts of coarse placer gold commonly sold by individuals to jewelers and specimen buyers are not reported to the U.S. Bureau of Mines or to the U.S. Bureau of Mint. Information about the age and type of lode deposit that was the source of the placer gold is discussed for each dis-trict.

A detailed search of the geologic and mining literature was made for information concerning all the placers. A list of literature references is given with each district; the annotation indicates the type of information found in each reference. Sources of information are detailed reports on mining districts, general geologic reports, Federal and State publications, and brief articles and news notes in mining journals. Two excellent gen-eral source books for the placer-mining districts of Nevada are Nevada Bureau of Mines Bulletins 18 and 30. (Smith and Vanderburg, 1932; Vanderburg, 1936a). County reports are also excellent sources for infor-mation relating to the mining districts. These source books give the loca-tion and history of most placer areas in the State. Many other publica-tions give additional information about the placers. A complete bibliog-raphy, given at the end of the paper, includes separate sections for all literature references and all geologic map references.

Map publications of the Geological Survey can be ordered from the U.S. Geological Survey, Distribution Section, Denver Federal Center, Denver, Colo. 80225; book publications, from the Superintendent of Documents, Government Printing Office, Washington, D.C. 20402.

CHURCHILL COUNTY

1. HOLY CROSS DISTRICT

Location: Terrill Mountains and Barnett Hills, T. 14 N., R. 29 E.; T. 15 N., R. 30 E.

Topographic map: Allen Springs 15-minute quadrangle.

Geologic map: Willden and Speed, 1968, Geologic map of Churchill County, Nevada (pl. 1), scale 1:200,000.

Access: From Fallon, 24 miles south on U.S. Highway 95 to dirt road leading east through Rawhide Flat. Mining areas are north and south of the main dirt road and accessible by minor dirt roads.

Extent: I have not found any information that locates the placer deposits in the Holy Cross district. The main part of the district is in the Terrill Mountains, west of the site of Camp Terrill on the south side of Rawhide Flat (secs. 10 and 11, T. 14 N., R. 29 E.), but the district also includes mines and prospects on the west side of the Barnett Hills (sec. 25, T. 15 N., R. 30 E.) about 6 miles northeast of the Terrill Mountains and on the north side of Rawhide Flat.

Production history: Placer gold was produced from the Holy Cross district in 1923 and 1947.

Source: The source of the placer gold was undoubtedly local material eroded from the veins in the Terrill Mountains and (or) Barnett Hills. The ore deposits in the Terrill Mountains (predominantly silver ore with smaller amounts of gold) are found in altered rhyolitic tuffs near dacite intrusives and are of Tertiary age. The ore deposits in the Barnett Hills are found in quartz veins in Cretaceous granodiorite; the Bimetal ores contain more free gold and silver than those in the Terrill Mountains.

Literature:

Vanderburg, 1940: Describes lode mines and prospects in Holy Cross district.

Willden and Speed, 1968: Describes bedrock geology and lode mines in Holy Cross district.

2. JESSUP DISTRICT

Location: Southern Trinity Range, T. 24 N., Rs. 27 and 28 E.

Topographic map: Desert Peak 15-minute quadrangle.

Geologic map: Willden and Speed, 1968, Geologic map of Churchill County, Nevada (pl. 1), scale 1:200,000.

Access: From Reno, about 66 miles east and north on U.S. Interstate 80 to the Jessup exit, a dirt road leading northwest about three-quarters of a mile to the townsite of Jessup in a pass through the Trinity Range.

Extent: A small placer deposit was discovered in 1932 on a low hill near the Valley King lode claims (location not recorded) in alluvium that consists of clay, small boulders, and well-rounded pebbles. Small lode deposits of gold, silver, and copper are found on the northern side of the range in secs. 8 and 17–20, T. 24 N., R. 28 E., and in sec. 22, T. 24 N., R. 27 E.

Production history: A few ounces of placer gold about 600 fine were dry-washed in 1932, but this production was not reported to the U.S. Bureau of Mines. In 1938, placer equipment was installed on the Valley King No. 3 placer claims, and 1 ounce of gold was recovered.

Source: The gold-silver deposits in the Jessup district occur in altered Tertiary volcanic rocks; the gold at the Valley King lode mine occurs in a free state, and the silver, in cerargyrite.

Literature:

Mining Journal, 1938a: Reports installation of placer equipment on Valley King No. 3 placer claims.

Vanderburg, 1940: Reports discovery of placer gold in 1932; production, and fineness of gold; lithology of alluvium.

3. SAND SPRINGS DISTRICT

Location: North end of the Sand Springs Range, T. 16 N., R. 32 E.

Topographic map: Reno 2 degree sheet, Army Map Service.

Geologic map: Nevada Bureau of Mines, 1962, Reconnaissance geologic map and sections, Sand Springs Range (pl. 4), scale 1:31,680.

Access: From Fallon, 28 miles west on U.S. Highway 50 to Summit King mine area, in Sand Springs Range, a quarter of a mile south of highway.

Extent: The placers are probably near the Summit King (formerly Dan Tucker) lode claims (approximately secs. 10 and 11, T. 16 N., R. 32 E.). These lode claims are the only productive ones in the district.

Production history: The placer gold was recovered in 1949.

Source: The Summit King group of claims are in quartz veins that trend east-west across the range and cut both metamorphic and volcanic country rocks. Free gold occurs in the quartz veins along with cerargyrite and argentite. The placer gold was probably recovered from material eroded from these quartz veins.

Literature:

Nevada Bureau of Mines, 1962: Describes ore deposits in Sand Springs Range; describes Summit King lode claims.

CLARK COUNTY

4. ELDORADO CANYON DISTRICT

Location: Along Eldorado Canyon, between Nelson and the Colorado River and adjacent banks of the Colorado River, T. 26 S., Rs. 64 and 65 E.

Topographic maps: Mount Perkins and Nelson 15-minute quadrangles.

Geologic map: Longwell, 1963, Geologic map and sections of area along Colorado River between Lake Mead and Davis Dam, Arizona-Nevada (pl. 1), scale 1:125,000.

Access: From Las Vegas, 31 miles southeast on U.S. Highway 95 to junction with State Highway 60, then east on State 60 12 miles to Nelson. State Highway 60 follows Techatticup Wash paralleling Eldorado Canyon on the north for 8 miles east from Nelson to the Colorado River.

Extent: Small amounts of placer gold have been found in the gravels of Eldorado Canyon, a major east-trending tributary of the Colorado River, and in bars along the Colorado River, for an unspecified distance south of Eldorado Canyon. The gold recovered from the bars along the river was fine grained and erratically distributed.

Production history: Recorded placer gold production attributed to the Eldorado Canyon district has been small; the bulk of the production was recovered between 1932 and 1933 from small-scale operations in bars along the Colorado River. Before the 1930's, little work was done on placer deposits in the area. Placer gold valued at $6,000 was recovered in the 1890's, but no further production was recorded until the 1930's. In 1909 a large dredge was constructed to work gravels on the Arizona side of the Colorado River opposite Eldorado Canyon. This operation was a failure. In recent years, very small amounts of placer gold have been recovered from the Eldorado Canyon placers, mostly from the Gresh Group of claims in the upper canyon near Nelson (sec. 4, T. 26 S., R. 64 E., Nelson quadrangle).

Source: The source of the placer gold is probably the lode mines near Nelson. The principal ores of the mines occur in fissures in Tertiary quartz monzonite. Much of the gold is contained in sulfides, and, except where the veins are oxidized, there is little free gold. The paucity of free gold and its small particle size may account for the limited placers in the area.

Literature:

U.S. Bureau of Mines, 1965: Placer-mining operations; placer claim named; production.

Vanderburg, 1936a: Extent; placer-mining operations; production.

5. GOLD BUTTE DISTRICT

Location: On the north side of Lake Mead in the hills south of the Virgin Mountains, T. 19 S., R. 70 E. (projected); T. 22 S., R. 69 E. (projected).

Topographic maps: All 15-minute quadrangles—Gold Butte, Iceberg Canyon, Virgin Basin.

Geologic maps: Longwell, Pampeyan, Bowyer, and Roberts, 1965, Geologic map of the Copper King and Gold Butte districts, Clark County, Nevada (pl. 12), scale approximately ½ in=1 mile.

Access: From Las Vegas, 74 miles northeast on U.S. Interstate 15 to

Riverside. From Riverside, light-duty roads lead south about 37 miles to Gold Butte area and about 15 miles farther south to Temple Bay area.

Extent: Placers are found in two areas in the isolated region north of Lake Mead and south of the Virgin Mountains. The Gold Butte district is in the vicinity of Gold Butte south of Tramps Ridge (T. 19 S., R. 70 E., projected, Gold Butte quadrangle). Temple Bar placer is on the north shore of Lake Mead, near Temple Bay (T. 22 S., R. 69 E., projected, Iceberg Canyon and Virgin Basin quadrangles).

The gravels at Gold Butte are 2–20 feet thick and contain fine gold, no larger than the size of a pinhead (about 1 mm), in the several feet of gravel immediately overlying bedrock. The gravels at Temple Bar and the surrounding area contain very fine gold associated with a large amount of black sand. Temple Bar was inundated by the rising waters of Lake Mead, but similar deposits farther inland from the original shoreline of the Colorado River (in the NE ¼ T. 22 S., R. 69 E.) still exist.

Production history: One ounce of placer gold is the only recorded production for the area, but reports of small-scale intermittent placer mining since 1926–27 suggest a greater actual production. One report indicates that value of some of the gravels at Gold Butte averaged $2.50 per cubic yard.

In the summer of 1967 the gravels around Gold Butte were placered with the aid of a steam shovel, and the Temple Bar area was being worked for both lode and placer gold, which were transported from the area by barges (P. M. Blacet, oral commun., 1967).

Source: The placer gold at Gold Butte was derived from erosion of the small gold veins found in the Precambrian granite that forms Gold Butte. The placer gold near Lake Mead might have been derived from similar veins.

Literature:

Longwell and others, 1965: Locates placer-mining operations; source of gold.

Vanderburg, 1936a: Location; history of placer-mining activity; extent of placer; depth of gravel; size and value of gold in gravels.

6. LAS VEGAS DISTRICT

Location: Along the former drainage of Las Vegas Wash, now inundated by Lake Mead, T. 21 S., R. 64 E.

Topographic map: Henderson 15-minute quadrangle.

Geologic map: Longwell, Pampeyan, Bowyer, and Roberts, 1965, Geologic map of Clark County, Nevada (pl. 1), scale 1:250,000.

Access: From Las Vegas, 16 miles southeast on U.S. Highway 95 to

Henderson; from Henderson, about 8 miles northeast on State Highway 41 to vicinity of Las Vegas Beach, south of Las Vegas Wash.

Extent: Unknown. A few reports indicate that gold occurred in gravels
along Las Vegas Wash, a tributary to the Colorado River, now inundated by Lake Mead.

Production history: Placer gold was produced from the Las Vegas district in 1905. In 1909 the Empire Gold Dredging Co. of Los Angeles
controlled 1,440 acres of placer ground in Las Vegas Wash on which
they planned to install a large dredge. These plans were not imple-
mented.

Source: Unknown.

Literature:

Mining World, 1909: Reports plans to dredge gold-bearing sands in
Las Vegas Wash.

7. SEARCHLIGHT DISTRICT

Location: West of the Colorado River on the east side of the Newberry
Mountains, Tps. 28–31 S., Rs. 65 and 66 E.

Topographic map: Spirit Mountain 15-minute quadrangle.

Geologic map: Longwell, 1963, Geologic map and sections of area along
Colorado River between Lake Mead and Davis Dam, Arizona-Nevada
(pl. 1), scale 1:125,000.

Access: From Las Vegas, 57 miles southeast on U.S. Highway 95 to
Searchlight. From Searchlight, light-duty roads lead east and south
about 16 miles to the north end of the Newberry Mountains. Dirt roads
lead farther south along the mountain flanks.

Extent: The lode-gold mines in the Searchlight district are in the low
hills between the southern Eldorado Mountains and the northern Newberry Mountains (T. 28 S., Rs. 63 and 64 E.), but the meager information available about the placers indicates that these deposits are
east of the main lode-gold mining area between the Newberry Mountains and the Colorado River (Tps. 28–31 S., Rs. 65 and 66 E.).
Most of the placer gold was recovered from gravels described as ancient
riverbed gravels, remnants of which occur in the vicinity of the Newberry Mountains. Some of the placer gold may have been recovered
from the main Searchlight mining area.

Production history: Most of the placer gold attributed to the Searchlight
district was recovered in the 1930's.

Source: Small gold-bearing quartz veins in Precambrian granite and metamorphic rocks have been mined in places in the Newberry Mountains.
Placer gold recovered in this area was probably derived from this type
of vein.

The ores in the Searchlight lode-mining district, which might have supplied some of the placer gold attributed to the district, are found in igneous rocks of Tertiary age. These deposits are largely oxidized and contained both gold and silver in the heavy-surface ores.

Literature:

Longwell and others, 1965: Describes geology and ore deposits (but not placer deposits) in the Searchlight district and Newberry Mountains.

Richardson, 1936: Describes methods used in testing gold-bearing gravels in Newberry Mountains; size of gravels; accessory minerals.

OTHER DISTRICTS

8. BOULDER DAM

Less than 1 ounce of placer gold was recovered from the vicinity of Hoover Dam (formerly Boulder Dam) in 1935, probably from sands along the Colorado River. In the early 1900's, several news notes in mining journals reported the existence of supposedly rich deposits of fine gold in Colorado River sands in this area, but none of these deposits were ever productive.

9. BUNKERVILLE DISTRICT

Placer gold is reported in sands and gravels in Nickel Creek (T. 15 S., R. 70 E.) and other creeks to the east of Nickel Creek, in the northern Virgin Mountains near Bunkerville Ridge. No placer production has been reported.

Literature:

Beal, 1965: Reports placer gold occurrences.

10. MUDDY RIVER PLACERS

In 1912 and 1932, small amounts of placer gold were recovered from the Logan and Muddy River (Moapa) districts. These districts (in T. 15 S., Rs. 66 and 67 E.) are along the Muddy River, a major tributary to the Colorado River. In 1931 very fine placer gold was discovered in bottom-land gravels along the Muddy River, 3 miles southwest of Moapa, but the deposit was considered to be uneconomic. The gold recovered from the Logan district (Logandale is 8 miles southeast of Moapa) probably came from the Muddy River gravels, although the Logan district may be synonymous with the Bunkerville or Gold Butte (20 miles east of Logandale).

Literature:

Nolan, 1936a: Reports placer bullion shipped from Logan district; questions location.

Vanderburg, 1936a: Describes placer deposits near Moapa; reports uneconomic concentration.

DOUGLAS COUNTY

11. MOUNT SIEGEL DISTRICT

Location: On the northeast flank of Mount Siegel in the Pine Nut Mountains, T. 12 N., R. 22 E.

Topographic maps: Mount Siegel and Wellington 15-minute quadrangles.

Geologic map: Moore, 1969, Geologic map of Lyon, Douglas, and Ormsby Counties, Nevada (pl. 1), scale 1:250,000.

Access: From Reno, 45 miles south on U.S. Highway 395 to Minden; from there, dirt roads parallel Buckeye Creek for 20 miles eastward to the Mount Siegel placer area in the Pine Nut Range.

Extent: Placers are found on the major headward fork of Buckeye Creek, which drains north along the east side of the crest of the Pine Nut Mountains. Some reports state that the placers are found over a large area covered by Tertiary sedimentary rocks, but practically all activity was concentrated at Slaters placer mine (NW¼ sec. 11, T. 12 N., R. 22 E., Wellington quadrangle) at an elevation of 7,100 feet.

The placers consist of unsorted debris composed of rocks, gravels, sands, and some large boulders; the gold occurs throughout the gravels and is both fine and coarse.

Production history: The Mount Siegel placers were discovered in 1891 and worked intermittently to 1943. In 1896 a company spent large sums on pipelines and tunnels to bring water from a small lake to the placers for use in hydraulic mining; lack of water resulted in failure of this operation. In 1896 the King brothers worked placers along bedrock, reportedly recovering $10 per day per man. Placer-mining activity was continuous to 1919. In 1903, 1904, and 1906 about 200 ounces of gold was recovered; in other years, the amount was less than 100 ounces.

Source: The placer gold is a reconcentration of material derived from Tertiary sedimentary rocks in recent gully gravels. No lode source is exposed in the Mount Siegel area; south of Mount Siegel, gold veins are found in quartz monzonite in the Silver Lake (Red Canyon) district. Presumably, erosion of similar veins during the Tertiary was the source of the placer gold that has been reconcentrated in the recent gravels on Buckeye Creek.

Literature:

Engineering and Mining Journal, 1896b: Reports developments at Buckeye placer; depth of gravel and production per day per man at King brothers workings.

Gray, 1951: Locates Slaters placer; age and derivation of placer gold; does not discuss lode source.

Lincoln, 1923: Location; history.

Moore, 1969: Locates Slaters placer mine; distribution of gold; production.

Overton, 1947: History; location; placer-mining operations in 1890's; distribution and size of gold in gravels; fineness of gold.

Smith and Vanderburg, 1932: Location; age of placer gravels; production for 1911; fineness of gold; estimate of total production (to 1932).

Vanderburg, 1936a: Location; early placer-mining operations; estimate of production; distribution of placers; size of gravel material; size and fineness of gold; source of the gold placers; problems in early placer-mining operations.

12. GENOA DISTRICT

Location: East side of the Carson Range in the Sierra Nevada, west of the Carson River, T. 13 N., R. 19 E.

Topographic map: Carson City 15-minute quadrangle.

Geologic map: Moore, 1969, Geologic map of Lyon, Douglas, and Ormsby Counties, Nevada (pl. 1), scale 1:250,000.

Access: From Reno, 35 miles south on U.S. Highway 395 to junction with State Highway 57; from there, about 8 miles west and south on State Highway 57 to Genoa. Placers are found west of town.

Extent: No descriptions of the exact location of the placers in the Genoa district are available. The placers are said to be in Tertiary gravels located a short distance west of the town of Genoa.

Production history: Minor gold and silver lodes were extensively prospected in the 1860's but were found to be of little economic value. Placer gold was recovered from the area in 1908 and 1909; the deposits were prospected again in 1916.

Source: The placer gold recovered from the district was material probably eroded from the small gold and silver lodes formed by intrusion of the Cretaceous granites into Triassic sedimentary rocks.

Literature:

Overton, 1947: Describes lode deposits in district.

Vanderburg, 1936a: Notes prospecting activity in 1916; states that Tertiary gravels are of doubtful economic importance.

OTHER DISTRICTS

13. MOUNTAIN HOUSE (PINE NUT) DISTRICT

A few ounces of placer gold have been recovered intermittently from placers in the southern Pine Nut Range in western Douglas County (T. 10 N., Rs. 21 and 22 E.). The source of the gold is unknown.

Literature:

Overton, 1947: Describes lode deposits of district.

ELKO COUNTY

14. CENTENNIAL (AURA, BULL RUN) DISTRICT

Location: Central part of the Bull Run Mountains, north of Bull Run Reservoir, Tps. 43 and 44 N., R. 52 E.

Topographic map: Bull Run 15-minute quadrangle.

Geologic map: Decker, 1962, Geologic map of the Bull Run quadrangle, Elko County, Nevada (pl. 1), scale 1:62,500.

Access: From Elko, 66 miles north and west on State Highway 11-43 to State Highway 11a; from there, it is about 12 miles north on 11a to the Centennial district.

Extent: Small placers are in the Centennial district along Blue Jacket Creek and Columbia Creek (T. 44 N., R. 52 E.) and in the Bull Run Basin (T. 43 N., R. 52 E.). The placers along Blue Jacket Creek are on the west side of the creek in sec 22 and at the junction of California Gulch and Blue Jacket Creek in sec. 26. The placers along Columbia Creek are north of the Columbia Ranch in secs. 24 and 25 where the creek drains southward. Placers in Bull Run Basin are along Sheridan Creek (secs. 3, 10, and 9, T. 43 N., R. 52 E.).

Production history: The placers in the Centennial district were found after the lode discovery in 1869 and were worked in the 1870's. I have found no estimates of this early production. In 1905 a company organized to work the gravels in Bull Run Basin spent considerable money on ditches, flumes, and pipes to develop the ground. The operation was abandoned after a few hundred yards of material was sluiced. Placer mining since 1905 has been sporadic.

Source: According to Decker (1962, p. 55), mesothermal vein deposits occur in limestones in the area near Blue Jacket Creek (formerly called the Aura district) and within and bordering Cretaceous intrusive stocks of dioritic composition in the area near Columbia Creek (formerly called the Columbia district). In contrast to the predominantly gold-sulfide veins of the Edgemont district on the west flank of the Bull Run Mountains, most of these veins contain silver with lesser amounts of gold and base metals. The Aura King group, in Blue Jacket Gulch, however, is located on a vein containing high gold values.

Literature:

Decker, 1962: Describes lode deposits.

Emmons, 1910: Reports failure of large-scale placer-mining operations.

Mining and Scientific Press, 1908a: Reports placer-mining activity in the valley of Sheridan Creek; states that the valley has some of the best placer ground in the area.

Mining World, 1907a: Number of acres of placer ground in Bull Run Basin; locates placer claims at Aura King group of lode claims.

Vanderburg, 1936a: History; lack of recent activity.

15. CHARLESTON (CORNWALL) DISTRICT

Location: In several gulches on the south side of Copper Mountain, east of the Bruneau River, Tps. 43 and 44 N., R. 47 E.

Topographic maps: Mount Velma 15-minute quadrangle; Wells 2-degree sheet, Army Map Service.

Geologic maps:

    Coash, 1967, Geologic map of the Mount Velma quadrangle, Elko County, Nevada (pl. 1), scale 1:62,500.

    Granger, Bell, Simmons, and Lee, 1957, Reconnaissance geologic map of Elko County, Nevada (pl. 1), scale 1:250,000.

Access: From Elko, 26 miles northeast on Interstate 80 to Deeth. About 8 miles north of Deeth, a dirt road leading from the Marys River Road crosses the low hills; about 38 miles north to Charleston.

Extent: Placer gold has been found in Badger, Union, and Seventy-Six Creeks, Pennsylvania Gulch (not shown on maps), and Dry Creek, all southwest-draining tributaries of the Bruneau River, and for an unknown distance along the Bruneau River in the vicinity of these creeks. The placer gravels are as much as 50 feet thick and consist of well rounded pebbles of volcanic rocks and smaller amounts of quartzite and granitic pebbles.

Production history: The placer deposits were the first metalliferous deposits discovered in the area and were presumed to have been found in 1876 because of the name, Seventy-Six Creek, given to the creek. Considerable placer gold was recovered from the gravels of Seventy-Six Creek during the years following discovery, but no production records, estimates of production, or discussions of the placers have been found in the literature. I estimate that probably about 300 ounces of placer gold was recovered before 1900.

    During the early years of placer-mining activity after 1876, Seventy-Six Creek was most actively worked, but during the 20th century mining has been concentrated in the other tributaries and along the river. In 1907 and in 1932, ambitious plans were made to mine placer gravels on a large scale in the vicinity of Badger Creek and the Bruneau River (secs. 22 and 27, T. 44 N., R. 57 E.); these plans were abandoned almost immediately.

Source: Schrader (1923, p. 83) considers the placer gold to have been derived by erosion of small gold veins in Paleozoic sedimentary rocks intruded by Cretaceous granitic rocks in the vicinity of Copper Mountain. The metallization is considered to be related to the Cretaceous intrusions and thus to be of Cretaceous age. Subsequent volcanic activity and erosion of the volcanic rocks probably caused mixing and diluting of the gold-bearing debris; the thick gravel deposits in the area are derived principally from the Tertiary volcanic rocks and contain only small amounts of quartzite and granitic debris from the metallized area.

Literature:

Coash, 1967: Notes Charleston as placer-mining district.

Lincoln, 1923: Location; history.

Schrader, 1923: Locates major placers; source.

Smith and Stoddard, 1932: History; gold yield per day per man in 1932; placer-mining developments at Earl Prunty Ranch.

Smith and Vanderburg, 1932: Names gold-bearing creeks; lithology and thickness of placer gravels; placer-mining operations in 1907 and 1932; average value of placer gravel at Prunty Ranch.

Vanderburg, 1936a: Virtually repeats Smith and Vanderburg, 1932; reports no development at Prunty Ranch placers; number of men working the placers in early 1930's.

16. COPE (MOUNTAIN CITY) DISTRICT

Location: On the south side of the Owyhee River, north of Mountain City and Sugarloaf Peak in the Bull Run Range (partly on Humboldt National Forest land), T. 46 N., R. 53 E.

Topographic map: Mountain City 15-minute quadrangle.

Geologic map: Coats, 1968a, Preliminary geologic map of the southwestern part of the Mountain City quadrangle, scale 1:20,000.

Access: From Elko, 87 miles north on State Highways 11 and 43 to Mountain City. Placers are on the south side of the Highway north of the town.

Extent: Small placer deposits occur along the Owyhee River, north of Mountain City and near Sugarloaf Peak. Grasshopper Gulch (secs. 26 and 35, T. 46 N., R. 53 E.), a north-trending tributary to the Owyhee River, is said to have been placered extensively in the early days of the district, discovered in 1869. Other deposits have been worked along banks of the Owyhee River for a few miles north of Mountain City.

Production history: No records of placer production directly credited to this area have been found. Any gold produced from the Owyhee River gravels and reported to the U.S. Bureau of Mines probably was included with production from the Van Duzer district on the south—even though the production was listed under the Mountain City or Cope district name. The placers were worked as early as 1870, when Chinese placer miners were reported to recover $2 to $3 per day per man from gravels on the north side of the river. The placers in Grasshopper Gulch were worked in the middle 1870's, for about half a mile along the gulch; considerable gold is said to have been recovered. I would estimate that not more than 200 ounces of placer gold was recovered before 1900 and perhaps another 50 ounces since.

Source: The source of the placer gold is small pyritic gold-silver veins in a granodiorite pluton. The age of mineralization is said to be Cretaceous.
Literature:
Raymond, 1872: Placer-mining activity on Owyhee River; yield per day.
Roberts and others, 1971: Source of placer gold; age of mineralization.
Smith, 1932: History; placer-mining operations on Owyhee River in 1932; source of gold.
Vanderburg, 1936a: Virtually repeats Smith and Vanderburg, 1932: names placer-bearing creeks; history and early production; placer-mining operations and developments in 1932.

17. GOLD CIRCLE (MIDAS) DISTRICT

Location: North of Squaw Valley between Midas Creek and Squaw Creek, T. 39 N., R. 46 E.
Topographic map: Midas 7½-minute quadrangle.
Geologic map: Rott, 1931, Geologic map of the Gold Circle mining district, Elko County, Nevada (pl. 1), scale 1 in.=1,000 feet.
Access: From Winnemucca, Humboldt County, 18 miles east on Interstate 80 to junction with State Highway 18; from the junction, Midas is 42 miles northeast on State Highway 18. Numerous light-duty and dirt roads lead from Midas to the mining area.
Extent: The location and extent of the placers are unknown. They are probably in the vicinity of the lode mines, in the low hills south and east of Midas (SW¼ T. 39 N., R. 46 E.).
Production history: The recorded production from the Gold Circle district occurred during the periods 1911–12 and 1920–21 and in 1941. The deposits and the mining methods were not described.
Source: The placer gold was probably eroded from the gold-silver veins that occur in volcanic rocks in the area. The ore deposits in the volcanic rocks formed about 15 m.y. (million years) ago (Miocene). Without accurate locations of the placers, it is impossible to ascertain which veins were the probable source.
Literature:
Rott, 1931: Describes lode mines in the district in detail.
Roberts and others, 1971: Dates mineralization in the district.

18. JARBIDGE DISTRICT

Location: Along the Jarbidge River, north of the Jarbidge Mountains. Tps. 45 and 47 N., R. 58 E.
Topographic map: Jarbidge 15-minute quadrangle.
Geologic map: Coats, 1964, Geologic map of the Jarbidge quadrangle, Nevada-Idaho (pl. 1), scale 1:62,500.
Access: From Elko, 26 miles northeast on Interstate 80 to Deeth. From Deeth, it is about 46 miles north on a dirt road to the site of Charleston.

A dirt road crosses the mountains east of Charleston to Jarbidge, a distance of about 16 miles; from Jarbidge, dirt roads parallel the river north of the town. Jarbidge is more easily accessible from Idaho.

Extent: Minor amounts of placer gold have been recovered from gravels along the Jarbidge River, especially north of the town of Jarbidge (T. 46 N., R. 58 E.). Residual soils adjacent to lode mines on the west side of the Jarbidge Mountains between Snowslide Gulch (sec. 3, T. 45 N., R. 58 E.) and Jack Gulch (sec. 4, T. 46 N., R. 58 E.) have yielded placer gold. The gold in the placers is very fine grained and has not formed economic concentrations—some reports state that it is so fine it floats away in water.

Production history: Placer production from the Jarbidge district has been very small, primarily because of the very fine size of the gold and difficulties in its recovery.

Source: The ore deposits of the Jarbidge district are gold-silver fissures and veins in Tertiary volcanic rocks, the probable source of the placer deposits.

Literature:

Buckley, 1911: Distribution, size, and shape of placer gold in the Jarbidge River; distribution of residual gold in soils; sketch map shows location of mines and placer claims.

Schrader, 1912: Notes absence of workable placer deposits; discusses distribution of fine gold in stream gravels; prospecting activity in 1910; number of colors of fine gold per pan; indicates areas where economic concentrations of gold might be found.

————1923: Notes presence of small placer gold deposits along Jarbidge River and East Fork; reasons for sparsity of placer deposits.

19. ISLAND MOUNTAIN (GOLD CREEK) DISTRICT

Location: Alluvial basin north of Island Mountain in the unnamed mountains between the Owyhee River and the Bruneau River, T. 44 N., Rs. 55 and 56 E.

Topographic map: Mount Velma 15-minute quadrangle.

Geologic map: Coash, 1967, Geologic map of the Mount Velma quadrangle, Elko County, Nevada, (pl. 1), scale 1:62,500.

Access: From Elko, 66 miles north on State Highways 11 and 43 to the Wild Horse Reservoir; from there, a dirt road leads east paralleling Penrod Creek for about 4 miles to Island Mountain and vicinity.

Extent: Placers occur in the alluvial basin on the north side of Island Mountain (N½ sec. 18, T. 44 N., R. 56 E.), and along Gold Creek, Hammond Canyon, and Coleman Canyon, which drain south into the basin (N½ T. 44 N., Rs. 55 and 56 E.). Most of the placer mining was in the shallow gravels of the alluvial basin, which extends about 1½ miles east-west and 1 mile north-south. The gravels were worked to an average depth of 7 feet, and much coarse gold, including nuggets valued at $50, was recovered.

Production history: The Island Mountain placers were discovered in 1873 by Emanuel Penrod, C. Rouselle, and W. Newton. Penrod and a few other miners worked the gravels by primitive methods for about 20 years before others became interested in the area. Penrod's claims occupied most of the alluvial basin. The early production is estimated at about $800,000, and those who worked the area reportedly recovered in gold as much as $1 per hour of labor on the gravels. Ambitious plans were made in 1897 to construct a ditch from the Sunflower Reservoir (built to store water for placer mining) to the placer deposits; but the ditch was not completed, and the boom placer-mining days of Island Mountain ended.

Placer mining was sporadic during the 20th century until 1934, when small-scale placer mining again became common. This activity lasted until the late 1950's.

Source: The placer gold is thought to have been derived from small vein and replacement lode-gold deposits in pre-Cenozoic rocks north of the alluvial basin and to have been transported south along Gold Creek and the creeks in Hammond Canyon and Coleman Canyon. The replacement deposits are associated with a small intrusive of probable Cretaceous age that is situated between Hammond Canyon and Coleman Canyon. The placer gold may have been derived from this area. Coash (1967, p. 19) states that much of the material in the placer gravels is similar to older prevolcanic gravels, and that the placer gravels, though postvolcanic (late Tertiary), may have been partly derived from reworking of the prevolcanic (early Tertiary) gravels.

Literature:

Burchard, 1883: Placer-mining operations; production in 1882.

Coash, 1967: Source and age of placer gravels; history and early production.

Engineering and Mining Journal, 1896a: Placer-mining developments by Gold Creek Mining Co.; high value of placer ground; average value.

————1897b: Test pit on Gold Creek near the mouth of Hope Gulch yielded $7.35 in gold from 5 cubic yards of gravel.

————1897c: Reports failure of Gold Creek Mining Co. to pay employees and states amount of attachments placed on the property.

————1898: Discusses reasons for failure of Gold Creek Mining Co.; reports resale of the property.

Lincoln, 1923: History.

Murbarger, 1957: Early history of discovery, production, and mining at Island Mountain.

Paher, 1970: History of the development of Gold Creek (Island Mountain); photographs of town, miners, and hydraulic mining.

Smith and Vanderburg, 1932: History of early placer-mining operations; early production per day per man; problems in placer mining; (name of placer creek erroneously given as Gold Run Creek instead of Gold Creek).

Vanderburg, 1936a: Early history; placer-mining operations during the period 1934–35; distribution of gold in gravels; size of gold recovered; source.

Whitehill [1875]: History of placer discovery and early operations; production.

20. TUSCARORA DISTRICT

Location: Southeast slope of Mount Blitzen in the Tuscarora Range, Tps. 39 and 40 N., R. 51 E.

Topographic maps: Tuscarora and Mount Blitzen 15-minute quadrangles.

Geologic maps:

Granger, Bell, Simmons, and Lee, 1957, Reconnaissance geologic map of Elko County, Nevada (pl. 1), scale 1:250,000.

Nolan, 1936, Sketch map of Tuscarora mining district, Elko County, Nevada (pl. 1), scale 1¾ in.=2,000 feet.

Access: From Elko, 45 miles northwest on State Highway 11 to junction with State Highway 18; from there, it is 10 miles west to Tuscarora on State Highway 18.

Extent: The placers in the Tuscarora district are largely confined to the low hills southwest of Tuscarora and north of McCann Creek (secs. 3 and 4, 9 and 10, T. 39 N., R. 51 E., Tuscarora quadrangle). Both hillside and gulch gravels were placered to depths of 4–10 feet. Other placers were worked on the south side of Beard Hill (sec. 7, T. 39 N., R. 51 E., Mount Blitzen quadrangle).

Production history: The Tuscarora placers were discovered in 1867 by the Beard brothers, Mr. McCan (sic), and Mr. Heath, prospectors from Austin. They found gold in small quantities for 3 miles along McCann Creek. News of the discovery reached Austin, and soon an influx of miners led to the discovery of placers and lodes in the hills north of the creek. Nuggets weighing 1 ounce were commonly found in the shallow gravel. Most of the early work was done by sluicing; to aid these operations, ditches were built to carry water 3–6 miles to the placers. The placer ground was turned over to Chinese miners in 1869, when the American miners began work on the silver mines to the north at Tuscarora. The Chinese placer miners reportedly recovered $2 to $15 per day per man. Early estimates of placer production stated that the placers yielded gold valued at about $7 million, but Nolan (1936b, p. 14), after studying production records, concluded that placer production did not exceed $700,000.

The placers have been worked only sporadically during the 20th century. The largest recovery was in 1902, 1905, and 1909; in other years, very small amounts of gold were recovered. The operation in 1909 was hydraulic mining by the Nevada Hydraulic Mining and Milling Co., which owned 480 acres of placer ground where the gravels ranged in depth from 4 to 15 feet. The value of the gravels was said to range from $1 to $3.50 per cubic yard, but the small amount of gold recovered indicates that the operation was not a financial success.

Source: The gold in the placers was derived from the small gold veins and stringers in bedded volcanic rocks in the area southwest of Tuscarora. The age of mineraliaztion is 38 m.y. (late Eocene or early Oligocene). The lode-gold deposits are confined to this area, as are the placers. Parts of the hillside gravels may have been residual concentrations of gold, as reports of the Nevada Hydraulic Mining and Milling Co. indicate that a gold ledge was found below placer gravels. The U.S. Geological Survey began (1970) studying this area in detail.

Literature:

Browne, 1868: Discovery history; early placer-mining operations; production; distribution of gold.

Emmons, 1910: History; production; distribution of gold; size of gold; early placer-mining operations.

Lincoln, 1923: History.

Martin, 1931: Production estimates; early history of placer discovery and mining.

Mining World, 1907b: Plans of Nevada Hydraulic Mining and Milling Co. to work placer ground; acres owned; depth of gravel; average value of gravel.

Nolan, 1936b: History; early production estimates (revises Emmons, 1910 estimate); locates placer claims; placer-mining operations in 1932; source of placer gold in that operation; source of placer gold in old placer operations.

Roberts and others, 1971: Age of mineralization in district.

Vanderburg, 1936a: Early history and production; extent of placered area; names richest gulches; depth of gravel; size of gold particles; size of largest nugget found; source; placer-mining operations during the period 1931–35; indicates potential dredging ground based on report by Emmons (1910).

Whitehill [1873]: Placer mining during the period 1869–72; number of miners; average yield per day per man.

21. VAN DUZER DISTRICT

Location: On the southwest side of the Owyhee River, south of Mountain City at the eastern edge of the Bull Run Mountains, Tps. 44 and 45 N., Rs. 53 and 54 E.

Topographic maps: All 15-minute quadrangles—Mountain City, Owyhee, Wild Horse.

Geologic maps:

Granger, Bell, Simmons, and Lee, 1957, Reconnaissance geologic map of Elko County, Nevada (pl. 1), scale 1:250,000.

Coats, 1968b, Preliminary geologic map of the Owyhee quadrangle, scale 1:31,680.

Access: From Elko, 81 miles north on State Highway 43 to Van Duzer Creek (6 miles south of Mountain City). Placers are along the creek, southwest of the Owyhee River.

Extent: Van Duzer Creek drains the low hills west of the Owyhee River for a distance of about 8–9 miles. Most placer-mining activity was concentrated in the upper reaches of the creek where the channel is narrow (secs. 32 and 34, T. 45 N., R. 53 E., Owyhee quadrangle). The gravel is generally less than 15 feet deep and consists of fine-sized granules and subangular and rounded pebbles. The gold varies from fine dust to nuggets weighing 5 or 6 ounces.

Production history: The Van Duzer placers were discovered in 1893 by Rutley M. Woodward, who worked the gravels by sluicing and hydraulic methods, reportedly recovering $50,000 in placer gold in the first years following discovery. The placers have been worked sporadically in the 20th century, generally with good returns.

In 1941 the Morrison-Knudsen Co. dredged the placer area from a point 1½ miles north of the junction with Cobb Creek (NE¼ sec. 32, T. 45 N., R. 53 E.) downstream about 1½ miles to farmland (SE¼ sec. 34, T. 45 N., R. 53 E.), recovering 2,128 ounces of gold. Subsequent operations by different companies in 1948 and 1949 on Van Duzer Creek and Cobb Creek recovered gold averaging 33 cents per cubic yard in 1948 and 70 cents per cubic yard in 1949. Most of the production credited to the "Mountain City or Cope" district by the U.S. Bureau of Mines was recovered from the Van Duzer Creek placers.

Source: The source of the placer gold in Van Duzer Creek is not well known. Small gold veins distributed throughout the Ordovician Valmy Formation probably are the source of the placer gold.

Literature:

Decker, 1962: States that placer gold deposits in Trail Creek and Van Duzer Creek (north of the Bull Run quadrangle) appear to be spatially related to an exposed thrust surface.

Emmons, 1910: Location; extent of placer; width and depth of placer channel; size of placer gold; source.

Lincoln, 1923: Location; history; geology.

Smith, 1932: History; early production; placer-mining operations in 1932.

Vanderburg, 1936a: Early production; size and fineness of placer gold; source; placer mining in 1932.

U.S. Bureau of Mines, 1941: Dredge operation on Van Duzer Creek described.

——1948: Placer operation on Van Duzer Creek; cubic yards mined; amount of gold recovered.

——1949: Placer operation at Estella Claim; cubic yards mined; amount of gold recovered.

OTHER DISTRICTS

22. ALDER (TENNESSEE GULCH) DISTRICT

Before 1900 placer deposits were worked along Gold Run Creek and Tennessee Gulch (T. 45 N., R. 56 E.) in the hills north of Island Mountain on the east side of Meadow Creek. No record of placer production exists.

Literature:

Engineering and Mining Journal, 1897a: Reports placer claims adjacent to Sunset group of lode claims in Tennessee Gulch.

Vanderburg, 1936a: Notes past placer-mining activity in Gold Run Creek.

23. GOLD BASIN (ROWLAND) DISTRICT

Placers were worked before 1900 and in 1931 along the north fork of the Bruneau River in the vicinity of Rowland (T. 47 N., R. 56 E.), but no production has been recorded from the area.

Literature:

Engineering and Mining Journal, 1893: Reports discovery of placers along Bruneau River.

Paher, 1970: States placer gold discovered on Bruneau River in 1869.

Vanderburg, 1936a: Reports small placer-mining activity in 1931; gravels valued at less than $1 per cubic yard.

ESMERALDA COUNTY

24. LIDA DISTRICT (TULE CANYON PLACERS)

Location: Along Tule Canyon, between Magruder Mountain and the northeast flank of the Sylvania Mountains, Tps. 6 and 7 S., Rs. 39 and 40 E.

Topographic map: Magruder Mountain 15-minute quadrangle.

Geologic map: McKee, 1968, Geologic map of the Magruder Mountain area, Nevada-California (pl. 1), scale 1:62,500.

Access: From Tonopah, 34 miles south on U.S. Highway 95 to Lida Junction; from there, 19 miles west to Lida on State Highway 3. About 5 miles west of Lida a dirt road branches off Highway 3 and leads south to Tule Canyon, which is paralleled by another dirt road.

Extent: Placer deposits occur along Tule Canyon and in some side gulches for an undetermined distance along the upper 4 miles of the canyon. Tule Canyon heads at the east side of an alluvial basin near Walker Springs and trends eastward between Magruder Mountain and the Sylvania Mountains to the alluvial valley at the east flank of the mountains, where it turns and drains south to Death Valley. Placers were worked in the alluvial basin near the head of the canyon (sec. '36, T. 6 S., R. 39 E.) and at other points along the eastward part of the wash. The alluvium along Tule Canyon is about 12–18 feet thick and is composed of sand, gravel, and medium-sized boulders in the upper part of the canyon; angular detritus containing large boulders are found in the lower part of the canyon.

Gold is concentrated along bedrock and within the overlying 6 feet of gravels above bedrock. The gold recovered at the Los Angeles Rock and Gravel Co. claims is coarse and rounded; gold recovered at the Ray White placer claims is coarse, angular, and attached to quartz.

Production history: The Tule Canyon placers are reported to have been discovered in 1876, but there is evidence of work by Mexicans before 1848. Early in the history of the district, the area was mined extensively by Chinese placer miners, who recovered large amounts of gold—estimated by some writers at more than $1 million. No real evidence exists to support this claim of high gold recoveries. During the 20th century, the placers have been worked continuously, although on a small scale. The operations of the Los Angeles Rock and Gravel Co. (1933–36) and those of E. E. Layton at the Ray White claims (1935) were two attempts at mining the gravels on a larger scale than usual in the district. The Los Angeles Rock and Gravel Co. worked an area 2 miles in extent along the upper part of the canyon, and E. E. Layton worked an area 1 mile in extent about 1½ miles below their operation. These operations recovered about 1,200 ounces of placer gold from 1933 to 1936. After cessation of these mechanized operations, placer mining was continued by individuals using drywashers or sluices utilizing water from the small springs in the area.

Source: The presumed source of the placer gold is gold ores in the Jurassic granites that form the Sylvania Mountains. Details of the gold occurrence are unknown.

Literature:

Stuart, 1909: Production estimate for Tule Canyon placers.

U.S. Bureau of Mines, 1935: Amount of gravel placered, in cubic yards; production.

Vanderburg, 1936a: History; location and extent of placers; placer-mining operations in 1930's; depth of gravel; placer-mining techniques; size of gravel material; size and fineness of gold; value of large nugget found; average value of gravel in lower Tule Canyon; source.

Whitehill [1875]: Placer-mining activity; size of nuggets (chispas) recovered.

25. SYLVANIA DISTRICT (PALMETTO WASH, PIGEON SPRINGS PLACERS)

Location: In the northwest end of the Sylvania Mountains between Leadville Canyon and Palmetto Wash, T. 6 S., R. 39 E.

Topographic map: Magruder Mountain 15-minute quadrangle.

Geologic map: McKee, 1968, Geologic map of the Magruder Mountain area, Nevada-California (pl. 1), scale 1:62,500.

Access: From Tonopah, 34 miles south on U.S. Highway 95 to Lida Junction; from there, 19 miles west to Lida on State Highway 3. The placer area is about 9 miles west of Lida on State Highway 3 in Palmetto Wash and in tributary washes reached by a dirt road leading south from the highway.

Extent: Small placer deposits occur on the east slope of the northwest end of the Sylvania Mountains in Leadville Canyon (sec. 20, T. 6 S., R. 39 E.) and in an unnamed major tributary to Palmetto Wash, near Pigeon Spring (secs. 16 and 17, T. 6 S., R. 39 E.).

Production history: The placers in the Sylvania district were reportedly discovered in 1866 and 1869 (Palmetto Wash placers). The gold found in Palmetto Wash contained large amounts of silver and was worth at that time only $13.50 per ounce. Activity in the area has been on a very small scale and intermittent.

Source: The placer gold probably comes from veins in Jurassic granites along the crest of the Sylvania Mountains. Silver and lead have been produced from these veins.

Literature:

Vanderburg, 1936a: Virtually repeats Smith and Vanderburg (1932). Brief note of placer occurrence in area called Pigeon Springs; methods of mining; year of greatest mining activity; placer-mining activity during the period 1929–30.

White, 1871: Discovery and location of Palmetto placers; size and fineness of placer gold.

OTHER DISTRICTS

26. DESERT (GILBERT) DISTRICT

A small amount of placer gold was recovered in 1935 and 1938 from this district, on the inner north slope of the Monte Cristo Range, a crescentic range in northern Esmeralda County (T. 4 N., R. 38 E., unsurveyed). The production in 1938 was recovered from the Gold Ridge claim. The placer gold was probably derived from late Tertiary gold veins in and near stocks of the Oddie Rhyolite.

Literature:

Ferguson, 1927: Describes lode mines.

Ferguson and others, 1953.

27. GOLDFIELD DISTRICT

Placer gold production was credited to the district for many years during the period 1909–49, but I have found no description of any placer deposit in this famous lode-gold mining district. The gold, which is very fine, occurs in late Tertiary veins that are generally not exposed at the surface. Placer gold was recovered in 1931 from erosional material from the Combination Fraction claims (sec. 1, T. 3 S., R. 42 E.). Some gold credited to placer production may have been recovered from mill tailings, which were extensively reworked.

It is also possible that placer gold production credited to the Goldfield district originated in outlying areas and was sold in, and therefore attributed to, Goldfield. One probable source for the placer gold is the Klondyke district (12 miles north of Goldfield). Placer deposits are reported for this district, but no production has been credited to it.

Literature:

Albers and Stewart, 1972.

28. GOLD MOUNTAIN (TOKOP) DISTRICT

This district, also known as the Oriental Wash district, covers a large area in southern Esmeralda County between Slate Ridge and Gold Mountain. (Tps. 7 and 8 S., Rs. 40 and 41 E., unsurveyed). The area has been prospected since 1866 and contains some high-grade gold veins such as that at the Oriental mine, discovered in 1871. The location of the small placer gold deposits reported by Vanderburg is unknown.

Literature:

Vanderburg, 1936a.

29. HORNSILVER DISTRICT

This district, formerly known as the Lime Point district, is in the area now called Gold Point (T. 7 S., R. 41 E.) on the south side of Lida Valley in southern Esmeralda County. Placers were discovered in 1946 and sampled in 1950 by a 45-foot shaft. Values reportedly increased in depth from $10.50 at the surface to $18.50 at the bottom of the shaft (Engineering and Mining Journal, 1950). However, no placer gold is known to have been recovered in 1950. In 1956, 16 ounces was produced from unlocated deposits, and none has been produced, or reported, since.

Literature:

Ransome, 1909a: Describes lode mines.

Engineering and Mining Journal, 1950: Reports developments in sampling activity by Goldpoint Mining Co.; gives results of sampling.

30. KLONDYKE DISTRICT

Although no placer gold production has been credited to this district, which is 12 miles north of Goldfield in T. 1 N., R. 42 E., a few reports indicate placer-mining activity on a small scale since the 1870's. A nugget valued at $1,200 was reported found. (See Goldfield district.)

Literature:

Vanderburg, 1936a.

EUREKA COUNTY

31. LYNN DISTRICT

Location: Along Lynn, Sheep, and Rodeo Creeks in the southern Tuscarora Mountains, T. 35 N., Rs. 50 and 51 E.

Topographic map: Rodeo Creek NE 7½-minute quadrangle (preliminary).

Geologic maps: Roberts, Montgomery, and Lehner, 1967, Geologic map of Eureka County, Nevada (pl. 3), scale 1:250,000; Map of part of the Lynn mining district, Eureka County, Nevada (pl. 7), scale 1 in.=200 feet.

Access: From Carlin, 20 miles northwest on light-duty road through the Maggie Creek Canyon, to Lynn Creek on the east slope of the Tuscarora Mountains.

Extent: Placers are found near the headwaters of Lynn, Sheep, and Rodeo Creeks on the summit of a low divide in the southern Tuscorora Mountains adjacent to the north end of Carlin gold mine. Most of the placers are concentrated on and near the crest of the divide (E½ sec. 11 and W½ sec. 12, T. 35 N., R. 50 E.) in streambed and hillside gravels. Lynn Creek was placered from near the headwaters downstream to the head of the dissected alluvial fan at the eastern edge of the mountain front (sec. 12, T. 35 N., R. 50 E.; sec. 18, T. 35 N., R. 51 E.). The gravels along this part of the creek have an average width of about 25 feet and a thickness of 10–28 feet; most of the placer gold was concentrated in the 4 feet of gravel immediately overlying bedrock. The hillsides on the south side of Lynn Creek, especially near the headwaters, were also placered; here, the gravels in one placer were 3–5 feet thick, and the gold was concentrated in the 1–1½ feet of gravel overlying bedrock.

Sheep Creek heads on the low divide a few hundred feet southwest of Lynn Creek and flows south and then west, into the Carlin mine tailings pond. The placers were concentrated in that part of the creekbed that trends south for a distance of about 2,500 feet (SE¼ sec. 11, T. 35 N., R. 50 E.). The gravels in this area occupy a channel 25–40 feet wide and about 8 feet deep; the gold is concentrated in the lowermost 1½–4 feet above bedrock.

Rodeo Creek drains northwest on the opposite side of the divide; the placers here were apparently concentrated in a small area near the upper reaches of the stream (NW¼ sec. 11, T. 35 N., R. 50 E.).

Production history: The placers in the Lynn district were discovered in 1907 by Joe Lynn; placer mining has been almost continual since that time and has yielded a few tens to hundreds of ounces of gold per year. The gold recovered from the Lynn district placers is exceedingly fine (average 920–960). Most of the placer mining was done by hand methods and small concentrating machines. Most of the placer areas along the upper reaches of the streams and adjacent hillsides that yielded as much as $1.50 to $8.00 per cubic yard have been thoroughly worked over, but the Southern Pacific Company (1964) reports an area of potential placer ground in the lower part of Lynn Creek (N½ sec. 17, T. 35 N., R. 50 E.) along a dissected alluvial fan. The company estimates reserves exceeding 1 million cubic yards, some of which carries 22 cents per cubic yard in placer gold.

Source: The placer gold in the Lynn district is derived from small auriferous quartz veins and stringers in chert of the Ordovician Vinini Formation that, in the Lynn area, contains shales and quartzites. Most of the placer-mining activity occurred in the upper reaches of stream channels that drain the mineralized chert. Only along Lynn Creek is gold found for an appreciable distance away from the bedrock source at the head of the creek.

The occurrence of gold in lodes and placers in the Lynn district contrasts with the finely disseminated gold in the carbonate rocks of the Silurian Roberts Mountains Formation about a mile south of the Lynn district at the Carlin gold mine. The Roberts Mountains thrust lies between the Carlin deposit and the placers of Sheep and Simon Creeks. The closeness of the Lynn and Carlin deposits suggests that the two ore bodies may be related despite their occurrence in such different host rocks. Current investigations by the U.S. Geological Survey on these ores may shed light on possible genetic relations of the two deposits.

Literature:

Emmons, 1910: Production from Hilltop placer claim; notes presence of placer gold in several gulches.

Koschmann and Bergendahl, 1968: Production.

Lincoln, 1923: Location, brief history, and extent of placers.

Roberts and others, 1967: Location and extent of placers; sketch map locates placers; width and depth of placer channel in Lynn Creek; width and depth of placer channel in Sheep Creek; source of placer gold.

Smith and Vanderburg, 1932: History; production estimates; accessory minerals; extent of placers; detailed description of placer-mining operations in Sheep and Lynn Creeks in 1932; depth and width of placer channel at Bulldog placer on Lynn Creek; thickness of paystreak; average gold content of gravel.

Southern Pacific Company, 1964: Locates potential placer ground in lower Lynn Creek; average per cubic yard; estimate of gravel reserves.

Vanderburg, 1936a: Detailed descriptions of placer-mining operations in 1935; location of placer operations; extent of placer ground that has been worked; size of nuggets recovered from Lynn Creek; placer extent on hillsides.

———1938a: Placer discovery; names placer creeks; distribution of gold in gravels; size and fineness of placer gold.

OTHER DISTRICTS

32. EUREKA DISTRICT

A small amount of placer gold was recovered from bench gravel at the north end of the town of Eureka (T. 19 N., R. 53 E) in 1937. The district is predominantly a silver district and probably does not contain appreciable amounts of placer gold. Production of 411 ounces of placer gold erroneously attributed to the Eureka district for 1941 was shown by the U.S. Bureau of Mines to have been actually recovered from the Lynn district.

Literature:

Nolan, 1962.

33. MAGGIE CREEK DISTRICT

Placers were reportedly worked near the headwaters of Maggie Creek (Tps. 35 and 36 N., R. 51 E.) between the Tuscarora Mountains and Independence Mountains. Maggie Creek receives drainage from Lynn Creek, a productive gold-bearing creek in the Tuscarora Mountains. No record of placer gold production has been found for the Maggie Creek district.

Literature:

Vanderburg, 1936a: States that placers were worked on Maggie Creek.

HUMBOLDT COUNTY

34. SAWTOOTH DISTRICT

Location: Northwest end of the Antelope Range, Tps. 34 and 35 N., Rs. 30 and 31 E. (unsurveyed).

Topographic map: Lovelock 2-degree sheet, Army Map Service.

Geologic maps:

Willden, 1964, Geologic map of Humboldt County, Nevada (pl. 1), scale 1:250,000.

Tatlock, 1969, Geologic map of Pershing County, Nevada, scale 1:200,-000.

Access: From Lovelock, 37 miles north on Interstate 80 to Imlay; from there, a light-duty road leads 28 miles northwest to Sawtooth Knob and placer area at the Humboldt County-Pershing County border.

Extent: The placer deposits in the Sawtooth district are in a nearly level area of approximately 6 square miles near the west side of Sawtooth Knob, a small prominence at the northern end of the Antelope Range. The gold is found in gravels composed of angular pebbles and few boulders. The gravels rest on a false bedrock of clay at depths of 8 inches to 2 feet; the clay in the gravels is said to be considerable and must be dried before drywashing the gravels. The gold recovered was coarse and rough and said to be 880 fine.

Production history: The placers were discovered in 1931 and were worked continually, although largely on a small scale, until 1942 and intermittently since that time. Some of the placer miners reportedly recovered $35 per day for short periods of time, but most of the miners were not so successful. Because there is little water in the area available for placer mining, dry concentration methods are used to recover the gold. Various attempts at large-scale placer mining were not successful.

Most of the production (810 oz.) was attributed to Humboldt County, where the major area of placer concentration is located, only 131 ounces of the total production being attributed to Pershing County. Placer production credited to the Sulphur district (located 6 miles west of Sawtooth Knob and noted for sulfur deposits) is included here in the Sawtooth district.

Source: Unknown.

Literature:

Smith and Vanderburg, 1932: Discovery of placer gold; placer-mining activity and production per day in 1931; character of placer gravel; distribution of gold in gravels; thickness of gravels; values recovered per cubic yard in 1932 from drywash operation; size of nuggets recovered.

Vanderburg, 1936a: Discovery and resulting placer-mining operations in 1931; extent of placer area; distribution of gold in gravels; size and fineness of gold; operations of Oregon-Nevada Mining Co.

——— 1936b: Discovery of placer gold; placer-mining operations; production; distribution of gold in gravels; size and fineness of placer gold.

——— 1938b: Discovery; production per day per man; extent of placer area; distribution of gold in gravel; size and fineness of placer gold.

35. DUTCH FLAT DISTRICT

Location: West flank of the Hot Springs Range, on the east side of Paradise Valley, T. 38 N., R. 40 E.

Topographic maps: Bliss and Osgood 15-minute quadrangles.

Geologic map: Hotz and Willden, 1964, Geologic map and sections of the Osgood Mountains quadrangle, Humboldt County, Nevada (pl. 1), scale 1:62,500.

Access: From Winnemucca, about 13 miles north on U.S. Highway 95 to junction with dirt road leading east across Paradise Valley to Sodarisi Canyon and Dutch Flat placers.

Extent: The placers of the Dutch Flat district occur in a small area about 1 mile from east to west and 2,000 feet from north to south, in the alluvium of Sodarisi Canyon and El Paso Gulch between Hot Springs Ridge and Belmont Hill, and at the head of the alluvial fan at the mouth of Sodarisi Canyon (sec. 17, T. 38 N., R. 40 E.). The placer deposits contain gold, scheelite, and cinnabar in economic quantities and occur in stream deposits along the gulches and in the alluvial fan, and in slope wash as high as 50 feet above the sides of the gulches. The stream deposits are 10–35 feet thick along the gulches; the slope-wash deposits are 5–25 feet thick. The alluvial-fan gravels are known to be at least 90 feet thick in some placers. The size, and possibly the concentration, of the various minerals change with distance from the source in the upstream end of the gulches. Assays based on the price of the metals in 1954 indicate an average value of $1.50 per yard of gravel in the lower stream gravels and alluvial fan. Values of placer gravels in the slope wash run higher, probably owing to the presence of large nuggets.

Production history: The Dutch Flat placers were discovered in 1893 and reportedly produced $75,000 in placer gold the first year after discovery. Estimates of total production of $100,000–$200,000 before the 1930's have been made, but recorded production figures do not support these estimates. However, it is possible that actual production was higher than recorded production, for before the 1930's the placer area was for many years leased to several individuals engaged in small-scale operations who might not have reported the amount of gold recovered. Lack of water for placer mining apparently has inhibited attempts at large-scale mining.

Gold recovered from the Dutch Flat area was frequently credited to the "Paradise Valley district" by Mineral Resources and Minerals Yearbooks. The Paradise Valley district proper is on the east side of the north end of the valley in the Santa Rosa Range, and, although some placer gold is said to have been recovered from the area, the discussions under "Mine descriptions" for most years of production indicate that the gold originated in the Dutch Flat district.

Source: The gold, scheelite, and cinnabar in the placers were derived from two, possibly three, different sources. The gold was derived from erosion of gold- and base-metal-bearing quartz veins that cut a small granodiorite stock (secs. 16 and 17, T. 38 N., R. 40 E.) of probable Cretaceous age and the Harmony Formation of Cambrian age; these veins also contain a few small grains of scheelite. The cinnabar occurs in fractures and interstices between mineral grains in Harmony sandstone, and is apparently slightly younger than the gold veins. Willden and Hotz (1955, p. 665) think that contact metamorphic deposits may be present in the district and may have contributed the scheelite (and accessory garnet) found in the placers.

Literature:

Hotz and Willden, 1964: Summarizes earlier report (1955) on gold-scheelite-cinnabar placer; adds detailed information about lode deposits that were source of placers.

Vanderburg, 1936a: History of placer mining in area; estimates of early production; methods of placer mining; extent and depth of placer gravels; size of large nugget; average fineness of gold.

——— 1938b: History; early production; average depth and value of gold-bearing gravel; placer operations in 1904; distribution, size, and fineness of placer gold.

Willden, 1964: Describes error in location of Paradise Valley district by earlier writers.

Willden and Hotz, 1955: Detailed description of three metals in placer deposits; extent of placer area, values of metals in placers; source of metals; variation in size of metals with distance from source.

36. GOLD RUN (ADELAIDE) DISTRICT

Location: East slope of the Sonoma Range, on the west side of the Pumpernickel Valley, T. 34 N., R. 40 E.

Topographic maps: Edna Mountain and Winnemucca 15-minute quadrangles; Gold Run 7½-minute quadrangle.

Geologic map: Gilluly, 1967, Geologic map of the Winnemucca quadrangle, Pershing and Humboldt Counties, Nevada, scale 1:62,500.

Access: From Winnemucca, about 16 miles east on Interstate 80 to light-duty road south of Golconda; from there, it is 10 miles south to placers in Gold Run Creek.

Extent: The placers of the Gold Run district are in the upper part of Gold Run Creek, just east of the town of Adelaide, in a basin composed of older fan gravels and recent alluvium (secs. 17 and 20, T. 34 N., R. 40 E.). The gravels in this area are about 10–14 feet thick, but gravels along lower Gold Run Creek are apparently as thick as 60 feet.

Production history: The Gold Run placers were reportedly discovered in 1886; early production is unknown. The placers were worked on a small scale throughout the 20th century, but except for the years 1903–04, when more than 1,000 ounces was recovered, total recorded yearly production has been small.

Source: The placer gold was probably derived by erosion of gold veins in the near vicinity of the placers. Such veins are exposed at the Crown mine, where gold and silver occur along a mineralized fault zone 10–80 feet wide between Cambrian and Ordovician shales and quartzites.

Literature:

Koschmann and Bergendahl, 1968: Placer production from Gold Run Creek.

Mining World, 1911: Reports plans to develop Gold Run placers; distribution of gold in gravels; thickness of gravels.

Southern Pacific Company, 1964: Locates area of potential placer ground along lower Gold Run Creek.

Vanderburg, 1936a: Early placer-mining activity; placer-mining operations at the Ontario-Nevada Mines Inc. property in 1935. Depth of gravel.

———— 1938b: History; production.

37. REBEL CREEK AND NATIONAL DISTRICTS

Location: West flank of the Santa Rosa Range between Santa Rosa Peak and Buckskin Mountain, Tps. 43 and 45 N., Rs. 38 and 39 E.

Topographic maps: Hinkey Summit and McDermitt 15-minute quadrangles.

Geologic map: Willden, 1964, Geologic map of Humboldt County, Nevada (pl. 1), scale 1:250,000.

Access: From Winnemucca, 49 miles north on U.S. Highway 95 to dirt road leading to Rebel Creek Ranch. Other roads parallel the main highway along the range front for 14 miles north to Canyon Creek.

Extent: Placers occur in several of the creeks that drain the west flank of the Santa Rosa Range on the east side of the Quinn River Valley. The creeks reported to contain placer gold are (from south to north) the Rebel, Willow, Pole, and Canyon. Most of the placer-mining activity was concentrated along Willow Creek (T. 44 N., R. 38 E.), which was worked as early as the 1870's by Chinese miners, who reportedly washed much coarse gold from 6 miles of the canyon. American miners turned the placer ground over to the Chinese miners because of the great depth of bedrock and the lack of drainage in the flat country-side. No descriptions of the exact location of the placers have been found.

Production history: Recorded placer production, which amounts to only a few ounces, has been credited to districts called Rebel Creek, National, and Quinn River. The gold credited to the National district was probably recovered from Canyon Creek (T. 45 N., R. 39 E.), as no placer gold is known to occur in creeks that drain the immediate vicinity of the National district (T. 46 N., R. 39 E.). The early production is unknown.

Source: The gold was probably derived from numerous minor gold-silver-copper-lead deposits of Late Cretaceous or early Tertiary age that are widely distributed throughout the range, or from the less abundant but richer gold-silver deposits of late Tertiary age that formed the ores in the Buckskin and National districts.

Literature:

Burchard, 1884: Reports placer mining in 1883.

Compton, in Willden, 1964, p. 122–127: Describes lode deposits in vicinity of placers.

Lindgren, 1915: States that no placers were found in vicinity of National district mines.

Paher, 1970: Placer-mining history in Rebel and Willow Creeks.

Vanderburg, 1936a: Names placer gulches; brief history of mining.

38. VARYVILLE DISTRICT (LEONARD CREEK PLACERS)

Location: On the west side of Leonard Creek Valley at the south end of the Pine Forest Range, north of the Black Rock Desert, T. 42 N., Rs. 28 and 29 E.

Topographic map: Duffer Peak 15-minute quadrangle.

Geologic maps:

Willden, 1964, Geologic map of Humboldt County, Nevada (pl. 1), scale 1:250,000.

Smith, 1972, Geologic map of the Duffer Peak quadrangle, Humboldt County, Nevada, scale 1:48,000.

Access: From Winnemucca, 33 miles north on U.S. Highway 95 to junction with State Highway 140; from there, 40 miles west on State Highway 140 to junction with dirt road leading 20 miles southwest to Leonard Creek Ranch. Dirt roads lead north into mountains and placer area.

Extent: Small placer deposits have been worked in gravels of various tributaries of Leonard Creek and in the gravels along the lower part of the creek, particularly in the area on the west side of the creek between Snow Creek on the north and the Pine Forest Range. But some men worked the gravels of Teepee Creek, a tributary on the east side of Leonard Creek. Most of the activity was concentrated in the vicinity of New York Canyon and the slopes and valley bottom of Leonard Creek (secs. 11, 14, and 23–25, T. 42 N., R. 28 E.).

Production history: Production from the Leonard Creek placers, which were discovered in 1914, has been small. Descriptions of the gravels and estimates of average value of gold per cubic yard seem to indicate that production was much lower than expected, or that actual production was rarely reported. Considerable attention was given in the late 1940's to placer gravels higher on the slopes of the Pine Forest Range than those worked near Leonard Creek; no production is known from the operations of the Eureka Hamburg Mining Co., which expended much effort in developing a placer area along the upper part of New York Canyon in a tributary locally called Fish Gulch. (secs. 13 and 14, T. 42 N., R. 28 E.).

Source: Unknown.

Literature:

Clark, 1947: Describes pilot plant for placer-mining gravels in Fish Gulch, tributary to New York Canyon; discusses topography of area where gold-bearing gravels are found; future plans of Eureka Hamburg Mining Co.

Vanderburg, 1936a: Repeats Smith and Vanderburg (1932). Brief history; placer-mining developments in 1931; average depth of gravel and average value of gravels; placer-mining activity in 1932.

—————— 1938b: History; production; placer-mining operations in 1937; width and depth of gold-bearing gravel channel; values per cubic yards; size and fineness of placer gold.

OTHER DISTRICTS

39. AWAKENING DISTRICT

The Awakening district, also known as the Amos district, is in the Slumbering Hills (Tps. 39 and 40 N., Rs. 35 and 36 E.) between Silver State Valley and Desert Valley. Placer gold was recovered from stream gravels along Teepee Creek in the Slumbering Hills in 1914. The Creek is not shown on the maps of the area but is probably on the east side of the hills that were the locale of prospecting activity at that time. One ounce of placer gold was recovered in 1937.

Literature:

Vanderburg, 1938b.

40. DUNNASHEE DISTRICT

The Dunnashee (or Donna Schee) placer area is on the east flank of the southern Jackson Mountains at Donna Schee spring (in sec. 25, T. 37 N., R. 31 E., or sec. 30, T. 37 N., R. 32 E., unsurveyed). Only a small amount of placer gold was recovered from gravels at the Last Chance placer mine.

Literature:

U.S. Bureau of Mines, 1954.

41. JACKSON CREEK DISTRICT

Placer deposits have been reported in the vicinity of Jackson Creek, a west-trending creek that drains part of the northern Jackson Mountains north of King Lear Peak. Iron mines occur near the headwaters of the creek (T. 40 N., R. 32 E.), but no descriptions of gold lodes or placers are available.

42. KINGS RIVER (DISASTER) DISTRICT

A few reports mention past placer-mining activity (1914, 1935) along China and Horse Creeks at the north end of the east side of the Kings River Valley (Tps. 46 and 47 N., Rs. 33 and 34 E.). Total production was probably very small.

Literature:

Lincoln, 1923.

Vanderburg, 1936a.

43. POTOSI DISTRICT

A few ounces of placer gold was recovered from stream gravels in the Potosi district (T. 38 N., R. 42 E.) on the east side of the Osgood Mountains. The gold was probably derived from erosion of the rich gold-silver ores of the Getchell mine and similar veins.

Literature:

U.S. Bureau of Mines, 1958.

Willden, 1964.

44. WARM SPRINGS DISTRICT

Placer gold was recovered in 1935 and 1941 from the Warm Springs district (Tps. 45 and 46 N., Rs. 28 and 29 E.) in the northern Pine Forest Range. The placer gold was probably derived from, and found near, the few small gold veins such as those at the Ashdown and Cherry Gulch mines in this area.

Literature:

Vanderburg, 1938b.

45. WINNEMUCCA DISTRICT

This district is a loosely defined district that includes the Blue Mountains, Krum Hills, and Winnemucca Mountain, north of the Humboldt River and south of the Slumbering Hills and Santa Rosa Range. Small gold-silver–base-metal veins occur over this wide area, and some of these veins have been worked as small lode mines. Most of the mines are found at various locations in T. 36 N., Rs. 35–38 E. The small intermittent production probably resulted from the work of individual miners placering gravels near some of the veins.

Literature:
Willden, 1964.

LANDER COUNTY

46. BATTLE MOUNTAIN DISTRICT

Location: North, east, and south slopes of Battle Mountain, south of the Humboldt River and northwest of the Reese River, Tps. 31 and 32 N., Rs. 43 and 44 E.

Topographic map: Antler Peak 15-minute quadrangle.

Geologic maps: Roberts and Arnold, 1965, Geologic map of the Antler Peak quadrangle, Humboldt and Lander Counties, Nevada (pl. 1), scale 1:62,500; Geologic map of the southeastern part of the Antler Peak quadrangle, Nevada, showing metamorphic zones and location of mines (pl. 3), scale 1:31,680; Copper Canyon fan (pl. 19), scale 1:9,600.

Access: State Highway 8a leads southwest from Battle Mountain (town) on U.S. Highway 40 to mining areas around the flanks of the Battle Mountain Range; light-duty and dirt roads lead from the main highway to placer areas.

Extent: Placers occur in many gulches that drain the north, east, and south slopes of Battle Mountain and in alluvial fans formed at the range front. The most productive placers are concentrated near two bedrock areas characterized by high copper-gold-silver metallization spatially and genetically related to centers of Tertiary intrusions. The placer area near the south flank of the range is bounded by Copper Canyon and Galena Canyon and includes placers in these drainages and in Box, Philadelphia, and Iron Canyons. The second placer area is in the Copper Basin lode area and includes Long Canyon and its tributaries, Licking Creek and Vail Canyon. Other placers are found on the north flank of the range, near small intrusive bodies at Snow and Piute Gulches, and in Elder Canyon. Some gold was recovered from Rocky, Willow, and Cottonwood Creeks outside the main placer areas.

In general, the gold is found in the lower gravel layers of alluvial fans, terrace gravels, and valley fill within the range ("older alluvium" of Quaternary age; Roberts, 1964b). This debris was eroded during an earlier, comparatively better watered, stage of the erosional cycle that shaped Battle Mountain.

The placers in the gulches occur in channels along bedrock and in terrace gravels and slope wash on the sides of the canyons; the main gold-bearing channels are buried by as much as 40–50 feet of relatively barren gravels or soils. Placers on the north side of Philadelphia Canyon occur in gravels overlain by basalts that are considered to be of late Tertiary or early Quaternary age.

The placers in the alluvial fan at the mouth of Copper Canyon (the most productive placer area) occur in lower gravel layers that are characterized by a higher degree of roundness, sorting, and washing than the overlying fan gravels, which are relatively unsorted, unwashed, and barren. Within these lower gravels, the gold occurs in channels, sheetlike bodies, and lenses having different concentrations in different horizons. Gold is found throughout most of the lower gravels (as much as 200 ft thick in the southern part of the fan), but a few gravel lenses are barren. The high-grade gravels occur in channels at bedrock near the head of the fan and in discrete lenses erratically distributed in the lower part of the fan. (See Roberts and Arnold, 1965, for detailed descriptions of gravel depth and value at individual placers.)

Production history: The lode deposits in the Battle Mountain area have been worked since 1866, but the placers apparently were not worked until 1909, when the discovery of rich gold lodes and associated placers near the mouth of Philadelphia Canyon stimulated a rush to this area, then called Bannock. Within 5 years, many of the other placers in the district were discovered and worked. Lack of water and the great depth of pay gravels were problems frequently encountered in small-scale mining of the placers. The early work was mainly drifting and slucing of deep gravels in Copper Canyon and other canyons in the district. The most productive periods were 1913–22, 1932–39, and 1947–55. Throughout these periods, the placers in Copper Canyon and in the Copper Canyon fan have been the most continuously worked and most productive deposits, but production from the other placers has been considerable. (See Roberts and Arnold 1965, for production data attributed to the different placers.)

The most productive operation in the district was that by the Natomas Co. (1947–55), which dredged a triangular area about 3,200 feet long and about 2,800 feet wide at the base, from the mouth of Copper Canyon down the fan (sec. 33, T. 31 N., R. 43 E.). The company used a dragline dredge to work the upper part of the fan between 1947 and 1948 and a large bucket-line dredge, capable of digging through 15 feet of gravel at the bottom of an 85-foot-deep pond, to work the downslope part of the fan between 1949 and 1955. Production attributed to this operation is partly confidential, but probably amounts to about 100,000 ounces in total.

Source: The placers in the Battle Mountain district are derived from lode deposits within the three areas of copper-gold-silver metallization present in the district. Gold is present in varying amounts in all the base-metal deposits. Metallization in the district is genetically associated with the intrusion of porphyritic quartz monzonite bodies, which at Copper Canyon are dated at 38 m.y. (Oligocene). The size, degree of roundness, and fineness, gold-silver ratio) of the placer gold and the lithology of the gold-bearing gravels indicate that the gold in the different gulches and alluvial fans was derived from lode sources close to the individual placers.

Literature:

Hill, 1915: Locates some placer workings; distribution of gold in gravels; richest areas of placer accumulation.

Huttl, 1950a: Details of dredge operation at Copper Canyon fan; depth of dredge pond and depth of gravel excavated; source of water and electricity for dredge.

Luther, 1950: Brief history of placer-mining activity at Copper Canyon placers. Detailed description of Natomas dredge operations at Copper Canyon; details of dredge construction and methods of placer mining.

Martin, 1910: Placer operations in 1910; size of placer gold recovered; describes lode mines at Bannock.

Mining World, 1940: Purchase price paid by Natomas to J. O. Greenan for placer area; depth of pay gravel; average value per cubic yard; plans for dredging.

Roberts, 1964b: Sections "Quaternary System" and "Geomorphology" pertain to origin and deposition of gravels that in many places contain economic concentrations of gold.

Roberts and Arnold, 1965: Detailed descriptions of placer areas at Battle Mountain. History of placer mining during the period 1910–55; scattered production data for selected placers; describes extent, lithology, depth, gold distribution, size and fineness, and average grade of individual placers; source of gold in gravels; describes details of lode mines from which placers have been derived.

Silberman and McKee, 1971: Dates intrusive rocks in Battle Mountain district.

Southern Pacific Company, 1964: Locates Copper Canyon placer; brief history of mining; average grade of gravel mined by bucket-line dredge during the period 1947–55; depth of high-grade gravels.

Theodore and Roberts, 1971: Reports trace-element concentration in placer gold from Iron Canyon placer and isotopic composition of lead contained in placer gold. Concludes that placer gold was derived from lode source within the zone of introduced pyrite.

Vanderburg, 1936a: Names placer gulches; early history of placer mining; estimate of placer production during the period 1910–35; describes Copper Canyon placers; size and fineness of gold recovered from gravels; details of placer-mining methods at individual placer properties at Copper Canyon in 1930's; source.

———— 1939: Describes depth, value, gold distribution, and size at Iron Canyon, Vail placer, and Copper Canyon deposits; summarizes placer-mining methods.

47. BULLION (TENABO) DISTRICT

Location: East flank of the Shoshone Range in the southern part of Crescent Valley, Tps. 28 and 29 N., R. 47 E.

Topographic map: Crescent Valley 15-minute quadrangle.

Geologic map: Gilluly and Gates, 1965, Geologic map of the northern Shoshone Range, Nevada, (pl. 1), scale 1:31,680.

Access: From Winnemucca, 85 miles east on U.S. Highway 40 to junction with State Highway 21; from there, 28 miles south on State Highway 21 to Tenabo. Placers are found south of Tenabo and 5 miles north in Mud Springs Gulch.

Extent: Placers are found in two areas in the Bullion district. The most productive deposits are in Mill and Triplett Gulches (secs. 8 and 9, 16 and 17, T. 28 N., R. 47 E.) south of the townsite of Tenabo on the eastern edge of the Shoshone Range. Small placers are found in various spots along Mud Spring Gulch and tributaries, Rosebud and Tub Spring Gulches (T. 29 N., R. 47 E.) between Granite Mountain and the eastern edge of the Shoshone Range.

The Triplett Gulch placers (secs. 16 and 17, T. 28 N., R. 47 E.) occur in shallow gravels, 1–12 feet thick, composed of sand, soil, and rock fragments less than 5 inches in diameter. The Mill Gulch placers (secs. 8 and 9, T. 28 N., R. 47 E.) occur in gravels composed of sand and medium-sized boulders, ranging in thickness from 10 to 45 feet above bedrock. Gold in both gulches, which head near the Gold Quartz mine (west edge of sec. 8), is both fine and coarse in size. Scheelite has been reported from the Mill Gulch placers.

The Mud Springs Gulch placers are found along bedrock in gravels 60–90 feet thick. The deeper parts of the gravels contain many small and large boulders and are cemented in places. The gold is coarser near the head of the gulch than in the lower parts. No descriptions have been found of the placers in the Rosebud and Tub Springs Gulches, tributary to Mud Springs Gulch.

Production history: Placers were first worked in 1907 in the Mud Springs area of the Bullion district. These small deposits were worked by small-scale hand methods and reportedly produced about $2,000 in placer gold. None of the placer production data from U.S. Bureau of Mines records can be directly attributed to the Mud Springs placer area.

The placers in Mill Gulch were discovered in 1916, but little is known of the development in this area until 1931, when small-scale operations began and continued until 1942.

The Mill Gulch Placer Mining Co. operated a dragline dredge and washing plant in Mill Gulch from May 1, 1937, to April 3, 1939. During this same period, many small operators worked placers in nearby Triplett Gulch and reportedly recovered substantial amounts of placer gold; therefore, the total amount of gold recovered by the dredge in Mill Gulch is difficult to determine. I estimate a production of about 6,000 ounces for the Mill Gulch Placer Mining Co. operation. In 1938 this company was the largest producer of placer gold in the State, recovering somewhat less than 4,200 ounces. In the last 3 months of operation in 1939, the dredge handled 101,382 cubic yards of gravel to yield 800 ounces of gold and 93 ounces of silver, an average of about 24 cents per cubic yard in placer gold.

The Triplett Gulch Mines, Inc., operated a nonfloating washing plant in Triplett Gulch that received gravels from bulldozers and carryalls from January 1, 1940, to December 15, 1940. The washing plant handled 120,000 cubic yards of gravel to yield 1,627 ounces of gold and 160 ounces of silver, having an average value of about 49 cents per cubic yard in placer gold.

Source: The gold in the Mill Gulch and Triplett Gulch placer deposits was derived from veins distributed near the margin of the granodiorite stock that underlies the central part of the Tenabo mining area. The veins occur both within the stock and in the adjacent chert and crop out mainly on the ridge between the two gulches. Both the stock and the gold mineralization are of Oligocene age. The placers in the Mud Springs area were probably derived from veins near the large mass of granodiorite at Granite Mountain.

Literature:

Silberman and others, 1969: Dates age of mineralization and host rock at Tenabo.

Southern Pacific Company, 1964: Locates areas of potential placer ground in Black Rock Canyon, Mud Springs Gulch, and Tub Springs, Gulch.

U.S. Bureau of Mines, 1938–41: Reports large-scale operations in Mill Gulch and Triplett Gulch; amount of gravel worked and ounces of gold recovered for some years.

Vanderburg, 1936a: Extent of placers; date of discovery; placer-mining operations during the period 1930–35; depth and value of placer gravels in different gulches; size of large nuggets recovered.

———— 1939: Placer discovery; placer-mining operations in Mill, Triplett, and Mud Springs Gulches.

Wrucke and others, 1968: Describes distribution of lode gold and associated elements in mining areas around Tenabo and Mud Springs.

48. HILLTOP DISTRICT (INCLUDING PITTSBURG)

Location: On the northeast slope of Mount Lewis in the northern Shoshone Range, T. 30 N., R. 46 E.

Topographic map: Mount Lewis 15-minute quadrangle.

Geologic map: Gilluly and Gates, 1965, Geologic map of the northern Shoshone Range, Nevada, (pl. 1), scale 1:31,680.

Access: From Winnemucca, 55 miles southeast to Battle Mountain on U.S. Highway 40; 2 miles past Battle Mountain, a light-duty road leads south about 10 miles to Rock Creek and Crum Canyon on the east side of the Reese River Valley. From the edge of the mountains, it is about 6 miles farther south to the headwaters of Rock Creek in Crum Canyon.

Extent: Gold-bearing gravels are found in the upper part of Crum Canyon (variously spelled Krum Canyon) near the junction of Maysville Canyon and Hilltop Canyon. The most actively worked placer is called the First Riffle claim and includes about 160 acres in Crum Canyon and Hilltop Canyon. The shaft shown on the topographic map as the Nelson mine (SW cor. sec. 28, T. 30 N., R. 46 E.) is probably one of the shafts in this placer group. In 1939 the property was explored by four shafts ranging in depth from 27 to 72 feet and one bedrock drift 500 feet long; the alluvium consisted of sand and mud near the surface and contained medium-sized boulders near bedrock where the gold was concentrated. A small placer was worked in 1937 near the Pittsburg mine (sec. 32, T. 30 N., R. 46 E.) on the slopes above Maysville Canyon.

Production history: Recorded production is small. John Nelson discovered placer gold in upper Crum Canyon in 1914, and lessees who worked the gravels between 1914 and 1916 are said to have recovered about $2,000 in placer gold by drift mining. This production was apparently not reported to the U.S. Bureau of Mines but has been added to the production table. Before Nelson's placer discovery, 75 ounces of placer gold was recovered from an underscribed deposit in the Hilltop district in 1911.

Source: The placer gold in the Hilltop district was probably derived from the free-gold-bearing fissure veins in altered cherts and quartzites in the Ordovician Valmy Formation. These veins are well developed in the mountain slopes between Maysville Canyon and Hilltop Canyon at the Red Top and Hilltop mines. The placer gold was probably transported down these canyons to their junction in Rock Creek in Crum Canyon.

Literature:

Vanderburg, 1939: Placer discovery; production; location of placer claims.

49. McCOY DISTRICT

Location: East flank of the Fish Creek Mountains, on the west side of the Reese River Valley, T. 28 N., R. 42 E. (unsurveyed).

Topographic map: McCoy 15-minute quadrangle.

Geologic map: Ferguson, Muller, and Roberts, 1951a, Geologic map of the Mount Moses quadrangle, Nevada, scale 1:125,000.

Access: From Winnemucca, 55 miles east on U.S. Highway 40 to Battle Mountain and junction with State Highway 8a. From there, it is 12 miles south on State Highway 8a to light-duty road leading southwest about 10 miles to Fish Creek Mountains. A dirt road leads 8 miles south to the McCoy district on the east side of the mountains.

Extent: Small placers were worked in the near vicinity of the Gold Dome mine, one of the group of mines northeast of the site of McCoy (approximately sec. 2, T. 28 N., R. 42 E., unsurveyed).

Production history: Small placer gold production has been recorded for 3 years, 1924, 1935, and 1939. The district was first located in 1914, and some unrecorded placer gold was probably obtained soon after.

Source: The placer gold was presumably derived from erosion of the lode-gold deposits in the area that contain free gold. The Gold Dome mine is a replacement deposit in diorite and contains abundant free gold. The placer gold recovered from the area in 1924 was 756 fine.

Literature:

Schrader, 1934: Notes past small-scale placer-mining activity near Gold Dome mine; describes geology and lode deposits.

U.S. Geological Survey, 1924: Fineness of placer gold recovered in McCoy district.

OTHER DISTRICTS

50. BIRCH CREEK PROSPECT

Birch Creek is on the eastern flank of the Toiyabe Range, southeast of Austin. Gold, silver, and lead occur in granitic rocks and adjacent metamorphic rocks of the area. Before 1938, shafts were sunk in the gravels of Birch Creek at the edge of the range (secs. 34 and 35, T. 18 N., R. 44 E.) to explore for placer gold; but the results were discouraging, and no recovery of placer gold was reported from the area.

Literature:
Vanderburg, 1939.
Stewart and McKee, 1968.

51. IOWA CANYON PROSPECT

Iowa Canyon drains the west slope of the Toiyabe Range, north of Mount Callaghan. Before 1936, placer gold was discovered in the canyon near the Joseph Phillips Ranch (unlocated). The same situation prevailed as at Steiner Canyon—a large flow of water was encountered when a shaft was sunk in an attempt to reach bedrock, and no placer gold was recovered from the area.

Literature:
Vanderburg, 1936a.

52. KINGSTON DISTRICT

This district is in Kingston Canyon, in the Toiyabe Range, south and east of Bunker Hill. The area has been prospected and mined for gold-silver deposits since 1863. A few ounces of placer gold was recovered from the Kingston claim (unlocated) in 1935.

53. REESE RIVER PLACER

The Reese River drains a large part of central Nevada from its head near the southeast summit of Toiyabe Dome in Nye County to where it joins the Humboldt River, east of Battle Mountain, in Lander County. Some placer gold was recovered from somewhere along this large drainage. Vanderburg (1939, table 7) includes the placer gold production with the production from the Reese River lode district at Austin, Lander County (T. 19 N., 44 E.). This district is predominantly a silver district, but minor amounts of gold have been recovered.

Literature:
Vanderburg, 1939.
Ross, 1953.

54. STEINER CANYON PROSPECT

Steiner Creek drains the west flank of the Simpson Park Range in eastern Lander County (Tps. 20 and 21 N., R. 46 E.). Placer gold was reportedly discovered in the 1870's by men digging a well to supply water for a stage station; but so much water was encountered at the depth of the placer gold that recovery was impossible. Again in the early 1930's, two attempts to dig to bedrock in the same well and in another shaft up canyon were given up soon after the water table was reached. So far no placer gold has been recovered from the canyon.

Literature:
Vanderburg, 1936a.

LINCOLN COUNTY

55. EAGLE VALLEY DISTRICT

Eagle Valley district (T. 2 N., Rs. 69 and 70 E.) is a small gold-silver-lead district between the Wilson Creek Range and the White Rock Mountains in eastern Nevada. The source of the placer gold recovered in 1935 is unknown.

Literature:

Tschanz and Pampeyan, 1970.

56. FREIBERG DISTRICT

The Freiberg district (T. 1 N., R. 57 E.) is at the north end of the Worthington Mountains, western Lincoln County. The district has a very small production of lode gold, and nothing is known about the placer gold production in 1935.

Literature:

Tschanz and Pampeyan, 1970.

LYON COUNTY

57. BUCKSKIN DISTRICT

Location: North end of Smith Valley between the Buckskin Range on the east and the Singatse Range on the west, T. 13 N., Rs. 23 and 24 E.

Topographic map: Como 15-minute quadrangle.

Geologic map: Moore, 1969, Geologic map of Lyon, Douglas, and Ormsby Counties, Nevada (pl. 1), scale 1:250,000.

Access: From Reno, 65 miles south and east on U.S. Highway 395 to Holbrook Junction and State Highway 3; 11 miles east on State Highway 3 to junction with State Highway 22. From there, it is about 18 miles north on State 22 to the north end of Smith Valley and placer area.

Extent: The placers in the Buckskin district are in the low hills that trend east-west between Lincoln Flat and Smith Valley. The placer area is in eastern Lyon County, and the major lode mines of the Buckskin district are in western Douglas County. Placer-mining activity was concentrated in the vicinity of low hills west of Wishart Hill (sec. 8, T. 13 N., R. 24 E.). The Ambassador placer (formerly Scott-Case) is in the upper part of the dry ravine between two low hills (NE¼ sec. 8), and the Guild-Bovard placer is on the alluvial fan below the dry ravine (southern part of sec. 8). Other placers are found west of these deposits along the southwestern base of Wishart Hill (mostly in sec. 9). The Ambassador placer consists of gravel composed of rhyolite fragments that range in depth from 2 feet in the upper part of the ravine to 16 feet at the lower part of the ravine on a rhyolite bedrock. The Guild-Bovard placer consists of well-cemented gravel, also largely composed of rhyolite fragments and lying on rhyolite bedrock, found at depths as great as 76.5 feet.

Production history: The recorded production from the Buckskin district placers is 75 ounces (for the years 1939–41), but some production (especially for the years 1933–35) may have been held confidential or may have been grouped with the Yerington district. Vanderburg (1936, p. 64) estimates a production of $9,000. The Ambassador Gold Mines, Ltd. operated the Ambassador placer during the period 1934–35. This company expended considerable money constructing a pipeline to facilitate hydraulic mining, but their enterprise failed. The Guild-Bovard placer was operated in 1933 and 1934 by a power shovel and portable washing plant.

Source: The gold may have been derived from the rhyolite bedrock (part of the Hartford Hill Rhyolite Tuff of early Miocene age). Overton (1947, p. 21), however, states that the placer gold was derived locally from erosion of gold- and copper-bearing ores that occur as contact-metamorphic deposits between Triassic sedimentary rocks intruded by Cretaceous granitic rocks.

Literature:

Engineering and Mining Journal, 1933a: Describes hydraulic operation by Ambassador Gold Mines, Ltd.

Nevada Mining Press, 1931b: Details of placer discovery by Fred Hughes; assays of samples given; average value of gold in clay material; extent of placers; names claims of placer ground.

———1931c: Reviews report on placers by Nevada Bureau of Mines (press release)—average value of gravel; extent; potential dredging ground indicated; source of gold said to be old river gravels 200 feet west of discovery ravine; extent of old river channel; size and shape of gravels; heavy accessory minerals in river channels compared with same in placer gravels.

Overton, 1947: Location; source.

Vanderburg, 1936a: History; production; description of discovery site and Scott-Case claims; placer-mining developments by Ambassador Gold Mines, Ltd.; placer-mining techniques; history of operation; size of gold recovered; depth of gravel; lithology of gravels; describes placer-mining operations at Guild-Bovard claim; techniques. Table gives depth of pay gravel and average value of pay gravel per cubic yard at Guild-Bovard placer; size of largest nugget found; fineness of gold.

58. YERINGTON DISTRICT

Location: In the Singatse Range between Gallagher and Mason passes at the north end of Smith Valley, T. 14 N., R. 24 E.

Topographic maps: Como and Wabuska 15-minute quadrangles.

Geologic maps:

Moore, 1969, Geologic map of Lyon, Douglas, and Ormsby Counties, Nevada (pl. 1), scale 1:250,000.

Knopf, 1918, Geologic map of the Yerington district, Nevada, (pl. 1), scale 1:24,000.

Access: From Reno, 35 miles east on U.S. Highway 40 to Fernley and junction with U.S. Highway 95A; from there, it is 34 miles south to Wabuska. About 5 miles south of Wabuska, dirt roads lead west to the Singatse Range. The dirt road 2 miles south of the Gallagher Pass road leads directly to placer area.

Extent: The placers in the Yerington district are in a stream channel that traverses the Singatse Range on the west side of Carson Hill (secs. 23 and 24, T. 14 N., R. 24 E., unsurveyed) 6 miles north of the Anaconda Co. copper mine at Weeds Heights. The stream channel is about 2 miles long and 300–600 feet wide; the gravels, which average 20 feet thick, were mined for a distance of about 0.9 mile along the channel between two parallel dirt roads that branch and rejoin at the margins of the placer area. This deposit is known as the Adams-Rice (or Guild) placer. The upper parts of the channel gravels consist of small sub-angular cobbles, pebbles, and sand, crudely or poorly stratified. In places, the lower parts of the channel consist of cemented gravels containing well-rounded boulders interpreted by some geologists as a part of a Tertiary river system. Most of the gold was apparently found in the recent gravels and was observed to have two shapes, angular and water-worn, indicating that some of the gold was reconcentrated from the older gravels.

Small amounts of placer gold were found in the alluvial fan at the mouth of the canyon (sec. 18, T. 14 N., R. 25 E., Wabuska quadrangle), but little work has been done in this area.

The Penrose placer is about 2 miles southwest of the Adams-Rice (or Guild) placer in Lincoln Flat. The exact location is uncertain but is probably in the low hills on the west side of the range and about 3 miles north of Wishart Hill at the north end of Smith Valley (approximately sec. 22, T. 14 N., R. 24 E., unsurveyed, Como quadrangle). The Penrose placer is in gravels of the Tertiary river channel, which are as much as 120 feet thick; the average value of the gravels is 20 cents per cubic yard. Penrose (1937, p. 4) states that 90 percent of the material in the 120-foot-deep shaft and drifts from the shaft was foreign to the area; he found well-rounded fragments of serpentine and pieces of steatite.

Production history: The placer production from the Yerington placers has been small, despite the active development of the placers during the 1930's. Some placer gold was reported from the district in 1917, but the Adams-Rice and Penrose placers were discovered in 1931 (1929–30 according to Penrose). The Adams-Rice placer was worked by a caterpillar-mounted dryland dredge, which reportedly recovered an average of 60 cents per cubic yard. The Penrose placer was worked by a dry-concentrating table.

Source: The ancient river channel that underlies the placer gravels in the Yerington district is thought to be a continuation of the Tertiary (early Miocene) fluvial conglomerate that occurs at the base of the Hartford Hill Rhyolite Tuff and overlies Cretaceous granodiorite, mapped by Knopf (1918) just south of the placer area. The conglomerate trends northeast from the north end of Wishart Hill. The conglomerate contains well-rounded, but unsorted and unstratified, cobbles and boulders; the cobbles are mainly andesite, and the boulders are granitic siliceous sediments. If some of the gold in the placers was derived by reworking this fluvial conglomerate, that gold was derived from a source exposed in the early Tertiary. The ultimate source of this gold is unknown.

The angular fragments of gold found in the recent gravels at the Adams-Rice (or Guild) placer are thought to be derived from quartz veins at the head of the canyon.

Literature:

Huttl, 1934: Describes dryland dredge used by Apex Mining Co.; details of gold-recovery techniques; average value of gravel mined.

Mining Journal, 1945: Reports development plans by Singatse syndicate to mine lava-capped placers.

Moore, 1969: Locates Guild placer mines.

Penrose, 1937: Describes Tertiary river channel; locates channel and traces extent for 6 miles; character of gold in channel; depth of shaft at Penrose placer; type of rocks found in shaft; difficulties in mining gravels.

Stoddard and Carpenter, 1950: Location of placer claims; placer-mining operations; production from these operations; depth of gravel; reported average value per cubic yard; size of gold recovered; source.

Vanderburg, 1936a: Describes Adams-Rice placer—location; average depth to bedrock; different types of gravel; lithology of ancient river channel gravel; size and shape of gold in gravels; size of largest nugget found; placer-mining operations. Describes Penrose placer—location; age of gravel channel; lithology of gravels; average value of gravels; size of gold.

59. SILVER CITY (GOLD CANYON, DAYTON) DISTRICT

Location: South slope of the Virginia Range, north of the Carson River, Tps. 16 and 17 N., R. 21 E.

Topographic maps: Dayton and Virginia City 15-minute quadrangles.

Geologic maps:

Moore, 1969, Geologic map of Lyon, Douglas, and Ormsby Counties, Nevada (pl. 1), scale 1:250,000.

Thompson, 1956, Geologic map and sections of the Virginia City quadrangle, Nevada (pl. 3), scale 1:62,500.

Access: From Reno, 29 miles south on U.S. Highway 395 to Carson City; from there about 11 miles northwest on U.S. Highway 50 to Dayton. Placers are west and north of Dayton and are accessible by dirt roads.

Extent: The first discovery of gold in Nevada was made in 1849 in gravels near the junction of Gold Canyon and the Carson River near the present location of Dayton. This discovery led, after 10 years, to the discovery of the Ophir mine of the Comstock Lode—the first major location on the lode. The Ophir was discovered by Peter O'Reiley and Patrick McLaughlin, two placer miners looking for unworked placer gravels at the head of Six Mile Canyon.

Gold and silver have been recovered from the gravels along Gold Canyon, Six Mile Canyon, and the north side of the Carson River. Early placer miners worked Gold Canyon from the vicinity of Gold Hill (sec. 32, T. 17 N., R. 21 E., Virginia City quadrangle, Storey County) downstream to the Carson River (sec. 23, T. 16 N., R. 21 E., Dayton quadrangle, Lyon County) and Six Mile Canyon (6 miles north of Gold Canyon in T. 17 N., R. 21 E., Virginia City quadrangle, Storey County) for an undetermined distance to its head. After the discovery of the Comstock lode, most placer-mining activity was concentrated in Gold Canyon, in particular, in the lower part of the canyon in the alluvial fan west of Dayton. This area, and the gravels in the townsite of Dayton, northwest of the Carson River, were the scene of large-scale placer-mining operations after 1940.

The gravels at the Rae placer (sec. 16, T. 16 N., R. 21 E.), the site of intense placer operations during the period 1920–23 and in 1940, are on a terrace sloping southwest to the Carson River. These gravels are at least 90 feet thick, but the gold values decrease below 40 feet. The gravels northwest of the Carson River within the townsite of Dayton (sec. 23, T. 16 N., R. 21 E.; the site of dredging during the period 1941–42) are more than 120 feet thick. The gold-bearing gravel there is overlain by barren soil 2–8 feet thick.

Production history: The Silver City district placers are the most productive in Lyon County and among the most productive in the State. Total early placer production is unknown, but placer mining probably continued on a small scale throughout the latter half of the 19th century. The most productive years were probably those between 1849 and 1859, before the discovery of the Comstock lode, when the area was solely a placer-mining region. Lord (1883, p. 24) estimates a placer production of $548,600 from 1850 to 1857.

Placer-mining activity during the 20th century has continued steadily, mostly on a small scale. Five large-scale operations since 1900 recovered considerable amounts of placer gold and silver. Gold Canyon Dredging Co. (1920–23) operated a large electric bucket-line dredge in Gold Canyon from about 2 miles south of Silver City (in stream gravels in the Manuel King placer ground) to an area west of Dayton (in terrace gravels of the Rae placer ground). This operation recovered a total of 14,621 ounces of gold and 7,482 ounces of silver. The Oro-Neva Dredging Co. (1940) operated a dragline dredge on the terrace gravels of the Rae placers, west of Dayton, recovering a total of 3,365 ounces of gold and 1,703 ounces of silver. The Dayton Dredging Co. (1941–42) operated a large dragline and floating washing plant on a strip of gravels 2,000 feet wide and 2,200 feet long on the north side of U.S. Highway 50 within the townsite of Dayton, close to the original placer discovery site. This operation recovered about 32,000 ounces of placer gold (silver recovery is undetermined). After World War II, the Dayton Dredging Co., then called the Grafe Dayton Dredging Co. (1946–47), resumed work on company placers, recovering about 3,900 ounces of gold. The Dayton Co. worked the placers again during the period 1952–54, but recovery was only about 500 ounces for each year.

Source: The placers in the Silver City district were derived from erosion of the Comstock lode. The lode deposits consist of brecciated quartz veins, in places exceedingly rich in silver sulfide and native gold (ranging in age from 13.7 to 12.4 m.y.; M. L. Silberman, oral commun., 1970) found at intervals along the Comstock fault and the Silver City fault. The ratio of silver to gold is reported as 40:1. The placer gold contains a large amount of silver (average gold fineness 660), and the weight of silver recovered with the placer gold amounted to about half that of gold recovered.

The possibility of finding placer gold in Tertiary gravel deposits in the area has been mentioned by many geologists. The young age of mineralization (Miocene or Pliocene) precludes the presence of placers in gravels older than Pliocene.

Literature:

Bonham, 1969: Summarizes earlier work on geology of the Comstock lode; dates mineralization of lode.

De Quille, Dan, 1891: Early history of placer discovery in Gold Canyon; production per day from gravels per man; production per day per man on residual gravels at Gold Hill; discovery by placer miners in Six Mile Canyon of the Ophir silver mine; production from this residual placer.

Gianella, 1936: Distribution of placer gold; ratio of silver to gold in ores and in placers.

Lincoln, 1923: History, placer-mining operations, and origin of the Gold Canyon placers.

Lord, 1883: Chapters 1–3 describe early placer mining; daily yield per day per man; early lode discoveries by placer miners; estimates of placer production; problems in placer mining.

Mining World, 1941a: Construction and operating details of 14-cubic-yard bucket dragline dredge used at Dayton, 1941–42; length, width, and depth of area to be dredged; distribution of gold in gravels; thickness of barren overburden.

Moore, 1969: Production during the period 1940–43; plate 2 locates placer deposits; brief history of early placer mining.

Paher, 1970: History of the town of Dayton includes brief history of placer mining; photograph of dredge used in 1897 included. Silver City and Johntown are described with history of placer mining near these sites.

Southern Pacific Company, 1964: Locates placer operation during the period 1920–23; reports extent of stripping on Southern Pacific land; concludes that part is worked out.

Stoddard and Carpenter, 1950: History and location of placer-mining operations during the period 1920–43; placer-mining techniques; production from the different operations; average value of gravel.

Thompson, 1956: Notes extent of placers; describes lode mines.

U.S. Bureau of Mines, 1940–43, 1946–47, 1952–54: Reports large-scale placer operations; production.

U.S. Geological Survey, 1920–23: Reports large-scale placer operations; production; yardage mined (1922).

Vanderburg, 1936a: History of early placer activity (1849–57); average wage per day per man; source; placer-mining operations during the period 1920–23; production; placer-mining techniques; describes Rae placer ground; depth of pay gravels; distribution of gold in gravels; average value of gravels; fineness of gold; history of operations along Carson River to recover gold in mill tailings from Comstock lode.

Young, 1920: Reports activities at opening day of dredge mining in Gold Canyon; summarizes early history of placering; gives construction features of dredge.

———1921: Detailed description and chronology of dredge construction; source of water and power used in dredge operations; brief description of placer ground and problems anticipated in dredging.

OTHER DISTRICTS
60. COMO (PALMYRA, INDIAN SPRINGS) DISTRICT

Placer deposits in the vicinity of the Como-Eureka mine (T. 15 N., R. 22 E.) in the northern Pine Nut Range were worked in 1883, but lack of water apparently inhibited development. No production directly attributed to this area has been recorded in the 20th century, but for many years production from this district was included with production from the Silver City district.

Literature:

Burchard, 1884, p. 537.

61. ELDORADO CANYON DISTRICT

Eldorado Canyon, a major tributary to the Carson River, forms the Lyon County-Ormsby County boundary. The district contains lignite deposits worked in the 1860's; no gold deposits have been described in the area. The small placer gold production recorded may have occurred near the junction of the canyon with the Carson River opposite the town of Dayton.

Literature:

Stoddard and Carpenter, 1950.

62. PINE GROVE DISTRICT

Only a few ounces of gold has been recovered from placer deposits in this district, which is on the east flank of the Pine Grove Hills (sometimes called the Smith Valley Range) in southern Lyon County. Placers occur on the slopes of Sugar Loaf Mountain near the mouth of Pine Grove Canyon (T. 10 N., R. 26 E.), but only a little work was done on these deposits.

Literature:

Vanderburg, 1936a. (Described under Mineral County.)

63. TALAPOOSA DISTRICT

Gold deposits have been known in the northern end of the Virginia Range since the 1860's, but little work has been done in the district. A small placer deposit is reported to have been worked in 1922.

Literature:

Stoddard and Carpenter, 1950.

MINERAL COUNTY
64. EAST WALKER DISTRICT (MOUNT GRANT PLACERS)

Location: On the west flank of the Wassuk Range, southwest of Mount Grant on land included in the Naval Ammunition Depot Reservation, T. 8 N., R. 28 E.

Topographic map: Mount Grant 15-minute quadrangle.

Geologic map: Ross, 1961, Geologic map of Mineral County, Nevada (pl. 2), scale 1:250,000.

Access: About 59 miles south of Fallon on U.S. Highway 95, a light-duty road paralleling Cottonwood Creek leads west and south through the Wassuk Range to the placer area. As access to the Naval Reservation is restricted, permission for entering may be required.

Extent: Placer gold is found in gravels of Lapon Meadow (variously spelled Laphan, Lapham Meadow) at the head of Lapon Creek (secs. 24 and 25, T. 8 N., R. 28 E.). The placer deposit worked by the Grant Mountain gold mine consisted of alluvium described as black loam overlying decomposed granite. (Previous studies have called this deposit volcanic ash.) The gold recovered from a channel 40 feet wide and 9 feet deep was coarse and had a fineness of 898.

Production history: The placers were first worked in 1906, when the area was withdrawn from the Schurz Indian Reservation; but early production is unknown and may have been included with production from the Pamlico area in the Hawthorne district. The placers were actively worked between 1935 and 1940 under the name Grant Mountain gold mine with a dragline mounted on caterpillars and a portable sluice. Gold recovery was reported to be 55 cents per cubic yard.

Source: Unknown. Small fissure veins in Cretaceous granites exposed in Cottonwood and Corey Canyons, north and east of Lapon Meadow, contain small amounts of gold. Similar veins may exist in the Lapon Meadow area but are either unexposed or not described.

Literature:

Vanderburg, 1936a: Brief description of placer deposits and mining operations during the period 1932–36; states that gold is found in bed of volcanic ash; size of large nugget recovered ($30). Reports discovery of placer gold in Baldwin Canyon south of the Lapon Meadow placer area. Placers described under the name "Hawthorne district."

————1937b: Placer-mining history at Lapon Meadow; placer-mining operations at Grant Mountain gold mine; estimate of early production; depth and width of placer channel; bedrock type; distribution, size, and fineness of gold; average value of gravels; placers described under the name "Mount Grant district."

65. HAWTHORNE DISTRICT (PALMICO PLACERS)

Location: Central western edge of the Garfield Hills on land outside the eastern boundary of the Naval Ammunition Depot Reservation, T. 7 N., R. 31 E.

Topographic map: Pamlico 7½-minute quadrangle.

Geologic maps: Archbold and Paul, 1970, Geologic map of the Pamlico mining district (pl. 1), scale 1:24,000; Geologic map of the main Pamlico mines (pl. 2), scale 1:2,400.

Access: From Hawthorne, a dirt road leads southeast about 10 miles to the Pamlico area.

Extent: Small placer deposits of unknown extent and exact location have been worked intermittently in the vicinity of Pamlico Hill and Pamlico Canyon (secs. 13 and 24, T. 7 N., R. 31 E.) at the western edge of the Garfield Hills.

Production history: The Pamlico placers have apparently been worked since 1908, the first year of recorded production attributed to the Hawthorne district. During 1915 and 1916, deep deposits at Pamlico in Pamlico Canyon were worked by drift mining, and slope wash or skree on Pamlico Hill was worked by placer methods. Subsequent placer mining has been on a small scale and intermittent.

Source: Gold-bearing quartz veins in volcanic rocks of the Excelsior Formation (Triassic?).

Literature:

Archbold and Paul, 1970: Brief history of placer mining; geology of bedrock and lode deposits.

Lincoln, 1923: States that placers were worked in the vicinity of the Pamlico mine by drift mining from 1915 to 1917.

Paher, 1970: Brief history of mining activity at Pamlico; photograph shows placer mine where a shaft was dug 170 feet to bedrock.

U.S. Geological Survey, 1915: Reports drift mining at Pamlico; method of mining; depth of gravels; average value of gold in gravels; production.

————1916: Reports placer mining of erosional material on Pamlico Hill; source.

Vanderburg, 1937b: Notes placer mining in Pamlico Canyon below the Pamlico mine; estimate of placer production in 1912; depth of gravels.

66. RAWHIDE (REGENT) DISTRICT

Location: At Hooligan and Balloon Hills, south of the Sand Springs Range and north of Alkali Flat, T. 13 N., R. 32 E.

Topographic map: Reno 2-degree sheet, Army Map Service.

Geologic map: Ross, 1961, Geologic map of Mineral County, Nevada (pl. 2), scale 1:250,000.

Access: From Fallon 22 miles south on U.S. Highway 95 to dirt roads leading east about 22 miles to Rawhide through Rawhide Flat.

Extent: Placers in the Rawhide district are found in Rawhide Wash and tributaries extending about 4 miles southeastward from the townsite of Rawhide to the alluvial fan at the base of the hills. Most of the placer-mining activity was confined to the main wash and side gulches between

Hooligan Hill on the west and Balloon Hill on the east. The gravels average about 15 feet deep on the southeast slope of Hooligan Hill but attain depths of 40–90 feet in the alluvial fan. The gold is erratically distributed in pay streaks that differ in thickness and depth throughout the lower part of the gravels. Within individual pay streaks, which are interpreted to be former creek channels, the gold is scattered or is locally concentrated. The erratic distribution of the gold is considered to be the result of erratic deposition during torrential floods of short duration.

Production history: The placer deposits at Rawhide were discovered and first worked at the time of the lode mining boom during the period 1907–8. Production to 1930 is estimated at about $200,000 to $250,000, although recorded production during this time amounts to only $34,800. Reports of high values of gold recovered from the placers in the early days support the high estimate of placer gold recovery.

The placer deposits of the Eagleville district northeast of Rawhide and 2½ miles east of State Highway 31 are probably part of this district. Vanderburg (1936a) states that a small amount of placer gold was recovered in 1906 and some placer gold was discovered in the canyon south of the Eagleville mine in 1931. These small deposits may be in either Churchill or Mineral County.

Source: The placer gold is derived from gold-silver quartz veins in altered Tertiary volcanic rocks that occur at Balloon, Murray, and Hooligan Hills.

Literature:

Mining and Scientific Press, 1908b: Notes presence of placer gold; mentions limited erosion and many cloudbursts, which should prevent finding large amounts of placer gold; size of gold recovered.

————1908c: Depth to bedrock in lower part of Rawhide Gulch.

Nevada Mining Press, 1930a: Placer-mining developments; reviews past placer-mining activity and production; reports value of gravel in 85-foot-deep shaft to bedrock; size of nuggets recovered; extent of placer ground; average depth of gravel.

————1930b: Average value of gravel at Hart shaft (see Nevada Mining Press, 1930a); reports sampling activity to determine possible dredgeable ground; extent of placer ground; methods of mining by early placer miners.

————1931a: Reports unexplained cessation of sampling by Idaho Gold Dredging Corp.; begun on March 29, 1931.

Paher, 1970: Locates Eagleville; small production during the period 1905–8.

Schrader, 1947: A comprehensive description of geology, lode, and placer deposits of Rawhide district; lithology of placer gravels; extent and distribution of gravels; thickness of gravels; distribution of gold in gravels; placer-mining history and operations; early production; methods of working gravels; problems in placer mining; bibliography of mining in district; describes individual placer claims.

Vanderburg, 1936a: Early history; estimate of production and yield per day per man; extent of profitable placer ground; depth of gravels on slope of Hooligan Hill and in alluvial fan; distribution of gold; size of large nugget; reason for discontinuing sampling operations in 1930.

———— 1937b: Extent of placers; location of most profitable placer diggings; depth of these richer gravels; location of deep gravels; concentration of gold; placer-mining history.

Wolcott, 1909: Extent of placer ground; details of drywashing machine used in placer mining; depth to bedrock in the lower part of Rawhide.

OTHER DISTRICTS

67. AURORA DISTRICT

A few ounces of placer gold (408 fine) was recovered from gravels in Bodie Creek (T. 5 N., R. 28 E.) in southwestern Mineral County during the period 1940–41. Bodie Creek receives drainage from both the Bodie district, Mono County, Calif., and the Aurora district, Mineral County, Nev. Although both districts are rich gold-silver lode mining districts, neither is noted for placer deposits.

68. BELL DISTRICT

This district is in the northern Cedar Mountains in southeastern Mineral County. Some placer gold was recovered from the Valley View claim (unlocated) in 1935. The gold was probably derived from erosion of the gold-silver quartz veins in volcanic rocks.

Literature:

Ross, 1961.

69. CANDELARIA (COLUMBUS) DISTRICT

I have found no information, other than production data, about the placer gold recovered in 1914 from this district, which is in the southern part of Mineral County (Tps. 3 and 4 N., R. 35 E.). The area is principally a silver district.

Literature:

Page, 1959.

70. SANTA FE DISTRICT

Placer gold was recovered in 1914 in this district, which is in the Gabbs Valley Range (T. 8 N., R. 35 E.) but also includes mines in the eastern Garfield Hills (T. 7 N., Rs. 32 and 33 E.). The ores of the district are most valued for copper, silver, and tungsten but gold is present in minor amounts.

Literature:

Ross, 1961.

71. SILVER STAR DISTRICT

Placer gold was recovered from the Big Dyke claim in the Silver Star district (Tps. 4–6 N., Rs. 32–34 E.). The district covers a large area in the Excelsior Mountains in southern Mineral County, and contains gold-silver veins with some tungsten. Free gold is found in brecciated and hydrothermally altered andesites in the eastern part of Camp Douglas area in the Silver Star district (T. 6 N., R. 34 E.).

Literature:

Ross, 1961.

72. TELEPHONE CANYON (PILOT MOUNTAINS) DISTRICT

Placer gold was recovered from deposits near the mouth of Telephone Canyon (near the Belleville mine, T. 6 N., R. 35 E.) in the Pilot Mountain Range in 1931. No record of this production exists, and little is known about the deposits.

Literature:

Vanderburg, 1936a.

NYE COUNTY

73. BULLFROG DISTRICT

Location: Bullfrog Hills and Bare Mountain on both sides of the Amargosa River. Tps. 10–13 S., Rs. 45–48 E.

Topographic maps: All 15-minute quadrangles—Bullfrog, Bare Mountain, Thirsty Canyon.

Geologic maps:

Cornwall and Kleinhampl, 1961, Geology of the Bare Mountain quadrangle, Nevada, scale 1:62,500.

————1964, Geologic map and sections of the Bullfrog quadrangle, Nye County, Nevada-California (pl. 1), scale 1:48,000.

Lipman, Quinlivan, Carr, and Anderson, 1966, Geologic map of the Thirsty Canyon SE quadrangle, Nye County, Nevada, scale 1:24,000.

Access: From Tonopah, 84 miles south on U.S. Highway 95 to Beatty. Mining areas are accessible by roads that extend from Beatty into surrounding hills.

Extent: Small amounts of gold have been recovered intermittently from placers in the Bullfrog district, which includes the Bullfrog Hills and the Amargosa Valley in the vicinity of Beatty (Tps. 11 and 12 S., Rs. 45 and 46 E., Bullfrog quadrangle). Some placer gold was reportedly recovered from gravels along the Amargosa Valley near the town of Beatty, east of the Bullfrog Hills.

Placer gold is reported to occur in the Carrara or Flourine district (sometimes also called the Beatty district) from the west slope of the Bare Mountains, southeast of the Bullfrog Hills (Tps. 12 and 13 S., R. 47 E., Bare Mountain quadrangle); the deposit was considered uneconomic.

One report notes a Paramount placer property located 15 miles northeast of Beatty in which the gravel is partly cemented and work was done by drifting into high-grade material. The location given for the property indicates the Camp Transvaal mining area (at the border of Tps. 10 and 11 S., R. 48 E., Thirsty Canyon quadrangle), southwest of Timber Mountain.

Production history: The placer gold production from this area has been small and intermittent. Most of the placer gold was recovered in 1912 and 1914.

Source: Since the exact location of the placers are not known, directly related source areas cannot be named. Numerous gold lodes present in both the Bullfrog Hills and Bare Mountain are the most probable source of the placer gold. In the Bullfrog Hills, the lodes are typically finely disseminated gold and silver in grains of pyrite in fissures and veins related to normal faults; in the Original Bullfrog lode (sec. 12, T. 12 S., R. 45 E.) native gold occurs as visible particles. On the west side of Bare Mountain, gold deposits occur in shear zones.

Literature:

Cornwall and Kleinhampl, 1964: Describes ore deposits.

Kral, 1951: Describes lode deposits at Bullfrog Hills and Bare Mountain; notes placer property northeast of Beatty; describes methods of working.

Vanderburg, 1936a: Notes placers in Amargosa River and Bare Mountain; states that the deposits have no economic importance.

74. JOHNNIE DISTRICT

Location: North end of the Pahrump Valley, in the low hills west of the Spring Mountains, Tps. 17 and 18 S., Rs. 52 and 53 E.

Topographic map: Mount Shader 7½-minute quadrangle (preliminary).

Geologic map: Cornwall, 1967, Preliminary geologic map of southern Nye County, Nevada, scale 1:250,000.

Access: From Las Vegas, 8 miles south on U.S. Highway 91 to junction with State Highway 16; from there, it is about 70 miles west and north on State Highway 16 to Johnnie and placer deposits north and south of the town.

Extent: Small placers are found in hillside and gulch gravels adjacent to, and below, many of the gold-quartz veins in the Johnnie district. Most placer-mining activity occurred in the gulches below the Congress mine (sec. 1, T. 18 S., R. 52 E.), located east of Mount Montgomery and half a mile south of the town of Johnnie. Most of the gold is concentrated in the 6 inches of gravel material that overlies bedrock and is overlain by as much as 25 feet of gravel. Samples of the 6 inches of material on bedrock have values as high as $6 to $30 per cubic yard, but the amount of material in this pay streak is unknown.

Placers are also found near the Johnnie and Overfield mines (sec. 20, T. 17 S., R. 53 E.) and the Labbe mine (sec. 30, T. 17 S., R. 53 E.) located northeast of the Johnnie on the west slope of the Spring Mountains. Parts of these placers are residual concentrations of gold, and parts are stream and hillside concentrations of transported gold.

Production history: The first recorded production of placer gold from the Johnnie district occurred in 1918, but the more productive accumulations of placer gold were discovered in 1920. This later discovery created considerable excitement and led to a short boom. A small amount of placer gold was produced almost yearly until 1950, and sporadically, until 1960. Most of the placer gold was recovered by drywashing the gravels. In 1949 the hillside below the Johnnie mine was mined by sluicing with water under high pressure. The amount of gold recovered by this technique did not differ appreciably from the amount recovered by drywashing techniques.

Source: The placer deposits were derived from the gold-quartz veins along faults in the Cambrian sedimentary rocks of the region thought to have formed mainly during the middle Cretaceous and to have remained active into the Tertiary.

Literature:

Cornwall, 1972: Notes placer-mining activity in 1949 and 1960; describes gold-quartz veins in district.

Engineering and Mining Journal, 1921: Names and locates placer leases in the Johnnie district.

Kral, 1951: History; location and extent of placers; type of placer; thickness of gravels; concentration and value of pay streaks; placer-mining operations.

Labbe, 1921: History; location; distribution of gold in gravels; thickness of pay streaks; characteristics of the gold; amount of gold in pyrite; placer-mining techniques with Mexican Air Jig.

Lincoln, 1923: History.

Vanderburg, 1936a: History; production; location and extent of placers; thickness of gravels and gold-bearing pay streak; size of large nugget; placer-mining activity in 1935.

75. CLOVERDALE DISTRICT

Location: Along Cloverdale Creek between the Shoshone Mountains and the Toiyable Range, north of Big Smoky Valley, Tps. 8–10 N., R. 40 E. (unsurveyed; on Toiyabe National Forest land).

Topographic map: Black Spring 15-minute quadrangle, U.S. Forest Service, scale 1:31,680.

Geologic map: Kleinhampl and Ziony, 1967, Preliminary geologic map of northern Nye County, Nevada, scale 1:200,000.

Access: From Tonopah, 2 miles west on U.S. Highway 6–95 to State Highway 89; from there, about 38 miles northwest to vicinity of Cloverdale Ranch; a dirt road paralleling Cloverdale Canyon leads north to placer area.

Extent: Placer gold occurs along the lower half of Cloverdale Canyon, which trends north-south for about 15 miles from its headwaters to Cloverdale Ranch. In 1906 the first placer was discovered 4 miles east of Cloverdale Ranch (approximately center of T. 8 N., R. 40 E.). Most, if not all, later work was concentrated in the gravels along Cloverdale Canyon for 8 miles north of Cloverdale Ranch (T. 9 N., R. 40 E.). The creek channel in this part of the canyon is 750 feet wide, and the depth to bedrock is 42–50 feet. Placers were worked in creek gravels and hillside gravels below East Golden, a small lode-mining area 8 miles north of Cloverdale Ranch, and at West Golden, on the west side of the ridge from East Golden. These placers and adjacent small lodes are probably in that part of the creek 2 miles below Farrington Ranch and 3 miles north of Four Mile Spring (secs. 8 and 17, T. 9 N., R. 40 E., unsurveyed).

Production history: Placer production from the Cloverdale district has been small and intermittent. Most of the placer gold was recovered by drywashing. Excessive water at the West Golden placer led to abandonment of placer mining. In 1928, ambitious plans were made to dredge 7½ miles of Cloverdale Canyon, after exploratory drilling indicated gold values of the gravels of 10 cents to $1.50 per cubic yard, but the plans were abandoned.

Source: The gold in the Cloverdale Canyon placers is probably derived by erosion of the lodes at East and West Golden mining areas. Kral, (1951, p. 44, 46) states that at East Golden the gold is found in shear zones in a brecciated rhyolite, and at West Golden (p. 44), in narrow veins.

Literature:

Kral, 1951: Locates two placer areas; placer operations in 1931 noted; problems in placer mining at one site; source.

Lincoln, 1923: Notes placer discovery.

Mining Journal, 1928: Reports plans to dredge Cloverdale Canyon; extent of placer area; length of placer area in canyon; width and depth of gravels in canyon; gold content of 10¢ to $1.50 per cubic yard.

Vanderburg, 1936a: Notes small-scale placer-mining activity beginning in 1906; reports plans to dredge area in 1931.

76. IONE (UNION) DISTRICT

Location: West slope of the Shoshone Mountains, on the east side of the Ione Valley, T. 13 N., R. 39 E. (unsurveyed); partly on Toiyabe National Forest land).

Topographic map: Ione 15-minute quadrangle.

Geologic maps:

Silberling, 1959, Geologic map and sections of pre-Tertiary rocks of the Union district, Nye County, Nevada (pl. 10), scale 1:24,000.

Vitaliano, 1963, Cenozoic geology and sections of the Ione quadrangle, Nye County, Nevada, scale 1:62,500.

Access: From Tonopah, 64 miles northwest to Ione on dirt roads that lead through Big Smoky Valley to Ione Valley. Placers are in immediate vicinity of Ione.

Extent: The placers in the Ione district occur in shallow gravels at the western edge of the mountains at two localities, 1 mile north and northwest of Ione (approximately secs. 28 and 29, T. 13 N., R. 39 E., projected) and 1 mile southwest of Ione (approximately secs. 2 and 3, T. 12 N., R. 39 E., projected). The gold is found in surface debris and gravels 1–2 feet thick overlying caliche layers adjacent to, and downslope from, small gold veins in the Tertiary volcanic rocks.

Production history: The Ione placers were first worked in 1909, and then intermittently until 1941. Total recorded placer gold production has been small, but the placers have attracted attention from companies considering large-scale operations. A group of Goldfield miners reportedly installed 2 miles of pipeline to bring water to the placers in 1909 in preparation for mining gravels valued at $1 per cubic yard. Between 1948 and about 1950, the placers north of the Ione road were sampled over an area 300 by 2,400 feet to a depth of 1 foot and were said to average $1.25 per cubic yard. In 1958 the Goldfield Rand Co. investigated a 1,280-acre placer claim in the Ione Valley and reported that drill samples of the gravels indicated a value of $1 per yard. This company had plans to install a large bucket dredge. None of the plans for large-scale operations materialized.

Source: Kral (1951, p. 196–197) states that weathering of high-grade gold stringers in Tertiary rhyolite are the source of the gold placers lying adjacent to and below these veins. One property where these gold deposits have eroded to form gold placers is the Bald Mountain Bill property, 1 mile northwest of Ione. The ore at this small property occurs in high-grade gold stringers and pockets in Tertiary rhyolite associated with jasper.

Literature:

Engineering and Mining Journal, 1958a: Reports drilling tests on 1,280-acre placer in Ione Valley by Goldfield Rand Co.; gravels said to indicate values of $1 per yard.

Kral, 1951: Locates placer deposits; reviews placer tests on different claims; value of gravels per cubic yard; source of placer gold.

Mining and Scientific Press, 1909: Reports developments at Ione placers; length of pipeline being built; value of gravels per cubic yard.

Smith and Vanderburg, 1932: Describes details of placer-mining operations in 1932; reports average value of gravel.

Vanderburg, 1936a: Brief description of placer-mining activity during the period 1932–35.

77. MILLETT (TWIN RIVER) DISTRICT

Location: East and west flanks of the Toiyabe Range at Ophir and Crane Canyons, T. 13 N., Rs. 41 and 42 E.

Topographic map: Round Mountain 30-minute quadrangle.

Geologic map: Ferguson and Cathcart, 1954, Geologic map of the Round Mountain quadrangle, Nevada, scale 1:125,000.

Access: From Tonopah, 65 miles north on State Highway 8a to dirt road leading west to Ophir Canyon; this road continues across the Toiyabe Range paralleling Ophir Creek and Clear Creek, half a mile south of Crane Canyon.

Extent: Small placers occur in gravels near the eastern range front along Ophir Canyon (approximately sec. 34, T. 13 N., R. 42 E.) and along Crane Creek, on the west side of the range (T. 13 N., R. 41 E.).

Production history: Placer gold was first credited to the district in 1910. The placers described here were located and prospected during the period 1946–47, but no production was recorded for 1946–47.

Source: Small gold veins occur at various localities throughout this area, but the major producing mine was the Murphy or Ophir, a silver mine in Ophir Canyon. Near the Murphy mine, high-grade gold ore occurs in a narrow quartz vein, and near-surface ores reportedly yielded appreciable quantities of gold. The placer gold was probably derived from this type of deposit.

Literature:

Kral, 1951: Locates small placers below mouth of Ophir Canyon.

Mining World, 1947: Reports plans of Natomas Co. to test gravels in Crane Creek; notes placer discovery the previous summer (1946) in Ophir Canyon in this same area.

78. MANHATTAN DISTRICT

Location: Southern end of the Toquima Range, on the east side of the Big Smoky Valley, T. 8 N., Rs. 43 and 44 E.

Topographic map: Manhattan and vicinity, special edition, scale 1:24,000.

Geologic map: Ferguson, 1917, Geologic map of the Manhattan district, Nevada, (pl. 6), scale 1:48,000.

Access: From Tonopah, 5 miles east on U.S. Highway 6 to junction with State Highway 8a; from there, 38 miles north to junction with State Highway 69, which leads 6 miles east to Manhattan mining area.

Extent: The placers in the Manhattan district were discovered in 1906 and have been the second most productive in the State during this century. These deposits are largely confined to Manhattan Gulch, which trends east-west across the west flank of the Toquima Range, draining the adjacent lode-gold mining area between Gold Hill and Palo Alto Hill.

Gold is found in four types of gravel corresponding to different stages of development of the gulch and adjacent hillsides. The oldest gravels are found in patches on both sides of the gulch at elevations of 20–70 feet above the present gulch level. These gravels are the remnants of an early stage in the development of the gulch before active downcutting of the canyon, and the gold concentration is much lower than that in the younger gravels. Active downcutting of the canyon resulted in the erosion of gold lode from Gold Hill and Palo Alto Hill, and deposition of gold-bearing gravels continued during downcutting of the canyon. These rich gold-bearing gravels are now seen as bench gravels on the canyon walls and as gravels in the deep channel of Manhattan Gulch. They overlie bedrock and are generally less than 10 feet thick, and are, in turn, overlain by as much as 40–100 feet of relatively barren overburden. These gravels are known to be Pleistocene in age on the basis of fossil remains recovered during placer operations. The youngest placers are found in recent wash on the hillsides, derived from erosion of the underlying veins.

The placers have been worked for a distance of about 6 miles from the vicinity of Gold Hill (approximately sec. 20, T. 8 N., R. 44 E., unsurveyed) to the eastern edge of the Big Smoky Valley (approximately sec. 21, T. 8 N., R. 43 E., unsurveyed). Detailed studies of the fineness of the placer gold recovered from the gravels (Ferguson, 1917, p. 191) show that the gold increases in fineness from east to west (range from 704 to 738) owing to solution of silver and base metals by the long action of ground water on the gravels, which have been undisturbed since the Pleistocene.

Production history: Since their discovery in 1906, the placers have been worked continuously by several methods in both large- and small-scale operations. Most of the placers in this district were reached by shafts and drifts and worked by sluices and drywashers. The numerous small operations had a significant yearly production. The most productive epoch in placer mining at Manhattan occurred between 1938 and 1946, when the Manhattan Gold Dredging Co., a subsidiary of Natomas Co. operated a floating bucket-line dredge which had 108 9½-cubic-foot buckets in Manhattan Gulch. The water for the operation was obtained from a 12-mile-long pipeline originating at Peavine. The dredge started at the eastern edge of the Big Smoky Valley (approximately sec. 21, T. 8 N., R. 43 E., unsurveyed) and worked eastward about 5 miles up the gulch to approximately the mouth of Black Mammoth Gulch (sec. 19, T. 8 N., R. 44 E., unsurveyed). Barren overburden averaged 30 feet deep. Both gulch gravels and bench gravels were dredged, the latter being pushed into the dredge path by large tractors, bulldozers, and scrapers. The company was given permission to operate at a reduced scale during wartime (1943–45). The dredge ceased operations at the end of 1946 and was later shipped to Copper Canyon, Lander County (see p. 37–38), where it was used in other large placer-mining operations.

Because mining took place after 1905, we have a reasonably complete record of the amount of placer gold recovered from the district. Recorded production indicates that $6,342,796 in placer gold was recovered between 1907 and 1967. The actual true placer production probably is not less than $7 million, as an unknown amount of placer gold produced was certainly not reported to the Government over this 60-year period. The dredge operation recovered somewhat less than 133,608 ounces of the total gold produced, and small-scale operations have recovered, over the entire 60-year period, at least 73,290 ounces.

Source: The placer gold in the Manhattan district was derived from the gold veins in the district, especially those at Gold Hill. The veins at Gold Hill occur in Cambrian limestones and schists of the Gold Hill Formation. Most of the placer gold is thought to be derived from the numerous, closely spaced quartz stringers in the schist rather than from the gold-bearing narrow fissures and replacement deposits in the limestone. The age of mineralization is thought to be early Tertiary or pre-Tertiary, the age of northwest-trending prevolcanic faults in the area.

Literature:

Clark, 1946: Describes techniques of placer mining used during large-scale operations; describes dredge, water supply, recovery methods; indicates extent of ground worked.

Ferguson, 1917: Detailed description of placers; includes history, production, mining operations; discusses development of present topography, distribution of gold in old gravels, deep gravels, and recent wash; size and fineness of placer gold; accessory minerals.

————1924: Brief account of placer occurrences summarized from Bulletin 640–J (1917), including maps, distribution of gold, character of gold, stream studies relating to size distributions of gold.

Ferguson and Cathcart, 1954: Text summarizes earlier study of Manhattan district (Ferguson, 1924).

Jones, 1909: Reviews geology of district; describes early placer-mining activity; lithology, thickness, and value of placer gravels; mining operations in 1909; includes claim maps of placer areas; average value of ground reported to be $10 per cubic yard.

Kral, 1951: Detailed description of history, geology, lode and placer mines of the district; much information taken from earlier writers. Adds information of production of placers to 1949, history of large-scale placer operations during the period 1938–46 and subsequent small-scale operations to 1949.

Martin, 1912: Production for 1911; depth and value of placer gravels; placer-mining techniques (paper similar to Toll, 1911).

Mining and Engineering World, 1913a: Describes drywash placer-mining plant owned by Thomas Wilson; extent and number of pay streaks in gulch worked.

Mining World, 1941b: Details of dredge and stripping operations in Manhattan Gulch; depth of barren overburden; total depth of dredgeable ground; details of stripping benches that dredge cannot reach; distribution of gold-bearing gravel; size of gravel; size of gold.

Paher, 1970: Photographs include many of early townsites and one of a dredge that operated near mouth of Manhattan Gulch; history of mining activity at Manhattan.

Stoneham, 1911: Describes rich placer gravels in main gulch below Manhattan discovered in April 1909; extent and thickness of gravels; placer-mining operations; drywash plant owned by Thomas Wilson described; value of gravels; presence of gold in dry lakebed in Smoky Valley.

Toll, 1911: Placer production for past year (1910–11); average thickness of gravel and pay streak; average value per yard in pay streak; methods of working placers.

U.S. Geological Survey, 1969: Relation of mineral deposits to faults and age of mineralization.

Vanderburg, 1936a: History and production of district; extent of placers; size of nuggets found; placer-mining techniques and methods to 1935; placer sampling by Natomas Co. described.

79. ROUND MOUNTAIN DISTRICT

Location: West flank of the Toquima Range, T. 10 N., R. 44 E.

Topographic maps: Round Mountain 30-minute quadrangle; Round Mountain #4, U.S. Forest Service, scale 1:31,680.

Geologic map: Ferguson and Cathcart, 1954, Geology of the Round Mountain quadrangle, Nevada, scale 1:125,000.

Access: From Tonopah, 5 miles east on U.S. Highway 6 to junction with State Highway 8a; from there, 49 miles north on State Highway 8a to State Highway 70, which leads 4 miles east to Round Mountain.

Extent: The Round Mountain placers, the most productive in the State during the 1900's, are remarkable for the large production yielded for the small area in which they are found. The main placers are on the south and west sides of Round Mountain, a small prominence about 6,800 feet in elevation at the western edge of the Toquima Range. Gravels on the east side of Round Mountain also have yielded some placer gold.

The main placers occur in coarse angular gravels and talus that are about 30 feet deep on the west edge of Round Mountain but thicken to more than 200 feet in the valley about three-quarters of a mile west of Round Mountain. In places, the deposits are considered to be residual, especially on the hillslope, but farther west towards Smoky Valley, the gold was probably transported several hundred feet. The gold is found throughout the gravels but occurs in highest concentrations on and near bedrock. The size of the gold is generally fine, only as large as a pinhead, but some nuggets also occur. The fineness of the gold is rather low, being only slightly finer than the gold in the adjacent lode deposits, which ranges from 574 to 696. The gold produced by the Round Mountain Gold Dredging Corp. during the period 1950–52 averaged about 630 fine.

Production history: The lode and placer deposits at Round Mountain were discovered in 1906. They were first worked by Thomas (Dry Wash) Wilson, who gained considerable fame for inventing and successfully operating drywashing machines at the Round Mountain placers. The placers have been intensely worked since their discovery and have yielded considerable amounts of placer gold almost every year until 1968. Throughout these 62 years, many different mining methods have been used to recover the gold. The earliest operations consisted chiefly of drywashing shallow gravels on the hillside, especially the claims held by Thomas Wilson adjacent to the Sunnyside lode. The drywash method reportedly recovered only about 70 percent of the placer gold; yet Wilson obtained gold worth $30,000 from these rich gravels in 70 days. Subsequent to initial work with drywashing machines, water was brought in and small-scale sluicing was used to mine some of the gravel. The sluicing led to the construction of a pipeline from Jefferson Canyon, about 3 miles away, and the commencement of hydraulic operations by the Round Mountain Hydraulic Co. Success of the hydraulic technique on shallow gravels despite the lack of water soon led to the construction of a

14-mile-long pipeline from Jett Canyon across Smoky Valley in the Toiyabe Range, where a better water supply was obtained. From 1906 to 1950 and during the period 1960–68, placer mining in Round Mountain was accomplished by all these methods (sluicing, hydraulicking, and drywashing) at various scales and intensity. The production from these small-scale operations amounted to about 87,000 ounces of the total estimated production for the district.

During the periods 1950–52 and 1958–59, two of the largest and most productive placer operations in the history of dry placer mining in the Southwest worked the Round Mountain gravels. The Round Mountain Gold Dredging Co. worked an area on the west side of Round Mountain (in secs. 19 and 30, T. 10 N., R. 44 E., unsurveyed) three-quarters of a mile west of the town of Round Mountain. The operations produced an open pit about 200 feet or more deep in the gravels, 4,000 feet long from north to south, and 1,800 feet wide from east to west. The first operation, 1950–52, utilized a dragline with scarifying plate located on the edge of the pit to break up the gravels and drop the debris into the pit, where a 7½-yard electric shovel delivered the gravels to a series of conveyors which, in turn, stacked the gravels for delivery by other conveyors to trommels to the mill. The mill in which the gravels were processed to remove the gold operated in the same manner as large floating dredges, but was stationary and had no dredge pond. A bucket-line dredge such as that used in Manhattan and Copper Canyon could not be used at Round Mountain because of the thickness of the gravels, large numbers of boulders, steep slope of the bedrock, and porosity of the gravels. The second operation, 1958–59, utilized a remodeled mill and electric shovels to break up the gravels and to strip the overburden, to mine selected areas within the pit. The placer gold production from these operations is estimated at about 145,000 ounces.

The Copper Range Exploration Co. and Ordrich Gold Reserves Co. renewed exploration of both placer and lode areas in the district during the period 1970–72.

Source: The lode deposits on Round Mountain that are the source of the placer gold consist of well-defined veins and numerous closely spaced small veins or stringers containing visible gold in quartz or associated with limonite and minor manganese oxide. These ores are found in a Tertiary welded tuff, formerly thought to be rhyolite.

Literature:

Engineering and Mining Journal, 1916: Hydraulic operations by Round Mountain Mining Co.; type of gravels mined; average value per cubic yard recovered in placer operations of previous season (1915?).

————1958b: Reports first phase of renewed large-scale operations at Round Mountain by Morrison-Knudsen Co.; amount of overburden stripped; amount of material to be processed; average value of material.

Ferguson, 1922: History; early production data; value of gravels from 0 to 60 feet; extent of gold-bearing gravels; lithology of gravels; size and fineness of gold recovered; source of placer gold; describes lodes from which placers were derived.

Ferguson and Cathcart, 1954: Text summarizes early study of Round Mountain district (Ferguson, 1922).

Huttl, 1950b: Describes placer-mining techniques at Round Mountain; location, depth, and size of placer gravels.

Kral, 1951: Detailed description of history, geology, lode and placer mines of the district; much information taken from earlier writers. Adds information on large-scale placer mining beginning at that time; describes general plan of operation.

Mining and Scientific Press, 1908d: Reports three hydraulic monitors recovering $1,000 per day in placer gold at Round Mountain.

Mining World, 1908: Reports production of $20,000 in fine gold and nuggets from first cleanup by Round Mountain Hydraulic Co.

————1950: Describes details of large-scale placer-mining operation; gives illustration of Pit Mining plan; describes talus mined by Round Mountain Dredging Corp.

————1951: Details of mining methods used by Round Mountain Gold Dredging Corp.; details of gold recovery techniques; size and fineness of gold recovered; includes flowsheet showing mining procedures.

————1959: Brief note which states that the "Round Mountain Gold Dredging Corp. is doing better in its second placering attempt than it did several years ago. It is now working selected areas by different mining methods."

Packard, 1907: Details of drywashing operations on Sunnyside claim; daily production; depth of gravel washed; compares drywashing at Round Mountain with drywashing at Manhattan.

————1908: Describes veins at Sunnyside claim believed to be partial source of placer gold; placer production by Thomas (Dry Wash) Wilson on Sunnyside placer claim; placer-mining developments by Round Mountain Hydraulic Mining Co.

Paher, 1970: Early history of mining activity at Round Mountain; photographs include drywashing activity; hydraulic mining; debris left by placer mining.

Ransome, 1909c: Distribution of gold-bearing gravel; placer-mining operations in 1908; production.

Tonopah Times-Bonanza, 1970: reports the start of evaluation programs at Round Mountain by Copper Range Exploration Co. and Ordrich Gold Reserves Co. Inc.

————1972: reports expansion of test program on lode and placer claims in Round Mountain District.

U.S. Bureau of Mines, 1950–52, 1958–60: Describes large placer-mining operations at Round Mountain.

Vanderburg, 1936a: Early placer-mining history and production; extent of placer; depth of gravel; size of debris in gravels; values for distribution of gold in gravels; details of different types of small-scale mining methods used in district.

OTHER DISTRICTS

80. BELMONT DISTRICT

Kral (1951) reports the presence of placer gold in Meadow and Antone Canyons (Tps. 9 and 10 N., R. 45 E.) in the northern part of the Belmont district on the east slope of the Toquima Range. Several groups of placer claims were located in these canyons, and nine holes reaching bedrock reportedly showed about 30 cents in gold per cubic yard. No production has been recorded from this district. The placer gold may have been derived from a base-metal deposit containing gold and silver in metamorphosed shales and limestones as at the War Eagle Group of claims in Antone Canyon.

Literature:

Kral, 1951: Locates placer claims; average value of gold per cubic yard.

81. CURRANT DISTRICT

An ounce of placer gold was credited to the Currant district in 1914. The few gold properties known in this small district are located on the south flank of the White Pine Range (T. 11 N., R. 59 E.) a few miles east of the town of Currant. The Sheperd property produced gold-lead-copper ore in 1914, and it is probable that the placer gold was recovered from the vicinity of this property.

Literature:

Lincoln, 1923.

Kral, 1951.

82. EDEN DISTRICT

Placer gold was recovered in 1935 from the Eden district, on the east flank of the Kawich Range, southeast of Kawich Peak near the headwaters of Eden Creek (T. 1 N., R. 50 E., Kawich Peak 15-minute quadrangle). The gold was probably recovered from debris eroded from gold veins or shear zones in rhyolite at the South Gold Mining Co. claims, at about 8,000 feet in the Kawich Range.

Literature:

Kral, 1951: Reports placer gold production from South Gold Mining Co. claims.

83. ELLENDALE DISTRICT

Placer gold was recovered from the Ellendale district in 1935. This small district is located in the hills of the southern Monitor Range east of Saulsbury Wash (variously spelled Salisbury Wash). Small mines scattered throughout this area (Tps. 2 and 3 N., Rs. 46 and 47 E.) were worked for gold in rhyolite.

Literature:

Kral, 1951.

84. FAIRPLAY DISTRICT

A small amount of placer gold (quantity confidential) was recovered in the late 1950's from small placers in the Fairplay district, which includes mines at the southern end of the Paradise Range (T. 10 N., R. 36 E.) near the abandoned mining town of Goldyke in western Nye County.

Literature:

Kral, 1951.

85. LODI (MAMMOTH) DISTRICT

Small placers were worked in the Lodi and Mammoth districts between 1935 and 1938. Lode-gold deposits in the districts (often considered to be one) are found in such widely separated areas as the southern tip of the Lodi Hills (T. 13 N., R. 36 E.) and the Ellsworth area on the east flank of the Paradise Range (T. 13 N., R. 38 E.) across the Ione Valley from Ione. The placer gold was probably recovered in the vicinity of the lode deposits.

Literature:

Kral, 1951.

86. LONGSTREET DISTRICT

Kral (1951) reports that placer gold has been recovered from gravels in Longstreet Canyon, on the east flank of the Monitor Range (T. 6 N., R. 47 E.). Three nuggets were reported found in the canyon, and some gold was recovered from surface detritus at the mouth of the canyon. The lode deposits of the Longstreet district consist of gold-silver ore in rocks described as rhyolitic tuff. There is no recorded placer production.

Literature:

Kral, 1951: Locates placers; describes recovery of nuggets.

87. TONOPAH DISTRICT

Some placer gold production is credited to this famous silver-gold mining district, in eastern Esmeralda County and western Nye County. I have found no descriptions of any placer occurrence.

Literature:

Kral, 1951.

PERSHING COUNTY

88. ANTELOPE (SCOSSA) DISTRICT

Location: Along the alluvial fan east and southeast of Majuba Hill and west of the Rye Patch Reservoir, T. 32 N., R. 32 E.

Topographic map: Lovelock 2-degree sheet, Army Map Service.

Geologic map: Tatlock, 1969, Preliminary geologic map of Pershing County, Nevada, scale 1:200,000.

Access: From Lovelock, 36 miles north on U.S. Highway 40 to light-duty road 1 mile south of Imlay that leads around the north end of the Rye Patch Reservoir to the junction with a dirt road leading south, a distance of about 12 miles. From the junction, it is about 8 miles south on the dirt road and 3 miles west to the placer area.

Extent: Placer are found in the alluvium at the east flank of the Antelope Range in the Majuba Hill area. The gold reportedly is close to the surface, and there is little overburden. Placer claims in T. 32 N., R. 32 E., located by the U.S. Bureau of Mines, are concentrated in sec. 8 (Majuba claim), sec. 20 (Rio Grande; Delta and Valley View), and sec. 30 (Delta and Valley View; Dice; Owens Circle).

Production history: The placers, reportedly discovered by Mr. C. E. Dice in July 1938, were most intensively worked from 1938 to 1941; more than 100 ounces per year was recovered during this period. The small placer production in 1934 and 1955 may have originated from the Scossa area of the Antelope district, which is northwest of Majuba Hill (T. 33 N., R. 30 E.).

Source: Unknown. The ore deposits of the Majuba Hill area are primarily silver, lead, tin, and copper. Some gold is associated with the base-metal deposits and might have been the source for the placer gold, but I have been unable to determine the source of the placers. In the Scossa district, 10 miles northwest of the Majuba placer area, gold veins occur in steeply dipping, easily eroded metamorphosed sedimentary rocks, a condition favorable for placer accumulation, according to Jones, Smith, and Stoddard (1931).

Literature:

Jones, Smith, and Stoddard, 1931: Describes gold veins and bedrock geology in district; notes characteristics of bedrock that create conditions favorable for catching eroded gold (at the time of survey, placers had not yet been worked).

Mining Journal, 1938b: Reports placer discovery by Charles E. Dice in Majuba area; reports gold is close to surface, with little overburden; states that drywashing methods are practical.

89. PLACERITES DISTRICT

Location: In the low hills southeast of the central Kamma Mountains and northeast of the north end of the Seven Troughs Range, at the corner of Tps. 32 and 33 N., Rs. 29 and 30 E.

Topographic map: Lovelock 2-degree sheet, Army Map Service.

Geologic map: Tatlock, 1969, Preliminary geologic map of Pershing County, Nevada, scale 1:200,000.

Access: From Lovelock, 18 miles west on State Highway 48 to junction with light-duty road leading north toward Sulphur along the east flank of the Seven Troughs Range. From this junction, it is about 39 miles north to Placerites area.

Extent: Placers in the Placerites district are in a small area of gravel hills adjacent to Rabbit Hole Creek on the northeast, about 8 miles southeast of Rabbit Hole. An unpublished map of T. 32 N., Rs. 29 and 30 E., and T. 33 N., R. 30 E., prepared by the Southern Pacific Company shows that most work was concentrated around the southern and eastern outer edges of the gravel hills. Placer gravels worked in small shallow ravines were 18 inches to 6 feet thick and rested on a bedrock composed of slates and shales. Most of the gold recovered is coarse.

Production history: The Placerites district was first worked in the early 1870's (some reports state 1850's) by "Mahogany Jack" and his three partners, who reportedly recovered $30,000 in gold. In the 1890's, placer miners hauled the gravels to Rabbit Hole Spring, about 9 miles northwest. Production during this time is unknown. The deposits were apparently not worked again until after 1928 and then intermittently until the present. The quantity of gold recovered from numerous small drywash operations each year has not been large. The Nevada-Montana Co. worked part of the company's 4,160 acres of claims in 1931 with a dragline scraper, but did not report the gold recovery. In 1969, Mr. Stanley held 28 placer claims in the district which were developed by an opencut 1,000 feet long and 25 feet wide and mostly from 6 to 10 feet deep, although the deepest part was 25 feet (R. C. Reeves, written commun., 1971).

Source: Unknown. The hills are composed of gravels possibly as old as late Tertiary (as mapped by Tatlock, 1969), and the placers were probably derived from these older gravels. The coarseness of the placer gold indicates primary derivation from a nearby bedrock source, possibly underlying the older gravels.

Literature:

Engineering and Mining Journal, 1931: Reports placer-mining operations by Nevada Montana Co.

Nevada Mining Press, 1929: Reports construction of reservoir by Ne-

vada Montana Mining Co.; discusses source of gold; states that no
trace of gold veins has been found in the vicinity of the placers.

Vanderburg, 1936a: Location; history and early production; placer-
mining activity and operations during the period 1928–35; depth of
gravel worked; size and fineness of gold; problems in placer mining.

————1936b: Early placer production; placer-mining development;
depth of gold-bearing gravel; fineness of gold; size of large nugget;
placer-mining operations in 1936.

90. ROSEBUD AND RABBIT HOLE DISTRICTS

Location: In the northern Kamma Mountains, north of Rosebud Canyon,
T. 34 N., Rs. 29 and 30 E. (projected).

Topographic map: Lovelock 2-degree sheet, Army Map Service.

Geologic map: Tatlock, 1969, Preliminary geologic map of Pershing
County, Nevada, scale 1:200,000.

Access: From Lovelock, 18 miles west on State Highway 48 to junction
with light-duty road leading north toward Sulphur along the east flank
of the Seven Troughs Range. From this junction, it is about 61 miles
north to Rosebud Canyon and placer area.

Extent: The extent and exact location of placers in the northern Kamma
Mountains are difficult to determine because of the lack of large-scale
maps showing the location of springs, gulches, and mines. Placers have
been worked northeast of Rosebud Peak in gravels on the side of the
mountains (approximately sec. 8, T. 34 N., R. 30 E., unsurveyed) and
on the southwest side of Rosebud Peak in gravels in ravines northwest
of Rosebud Canyon (S½ T. 34 N., R. 29 E., unsurveyed). Other placers
were worked in ravines known as Coarse Gold Canyon, Red Gulch,
Long Gulch, and Barrel Springs Canyon, said to be tributaries of Rosebud Canyon.

Most of the gravels in the ravines where gold was found ranged in
thickness from 2 to 12 feet. The gold-bearing gravels characteristically
overlie a false bedrock of clay; shafts sunk below this horizon yielded
little gold. The gold particles recovered from the area are flat; much
of the gold ranges in size from particles having a weight value of a few
cents to a few dollars.

Production history: Although most descriptions of placer mining here state
that work was concentrated in the Rabbit Hole district (west side of
the Kamma Mountains), all but 89 ounces of production was credited
to the Rosebud district (east side of the Kamma Mountains). The
placers near Rosebud Canyon were reportedly first worked in the 1870's
by Chinese placer miners, who recovered several thousand dollars in
placer gold, but placer mining before the 1930's was intermittent. From
1933 until 1942, the placers were worked continuously, by individuals

who drywashed the gravels and by companies who used dry-concentrating plants, power shovels, and trucks to mine the gravels. The largest yearly production was for 1939 and 1940, when placer-mining operations on the Janke group of claims in Barrel Springs Canyon and on the Rio Seco claims (unlocated), produced a total of 3,008 ounces of gold. Production continued intermittently from 1943 to 1963. Most of the placer ground is controlled by Constant Minerals Separation Co., who have operated placers in the district since the 1940's.

Source: The rocks exposed in the northern Kamma Mountains consist of Tertiary volcanic and sedimentary rocks in the central part and on the east flank, and of Tertiary sedimentary rocks, Tertiary and Quaternary gravels, and Quaternary alluvium on the west flank. The Janke claims were reportedly in ancient lakebed gravels. The placer gold was probably derived by erosion of gold-silver veins, in the northern Kamma Mountains, such as those at the Brown Palace mine, but, because there have been no detailed studies of the area, this derivation is not certain.

Literature:

Engineering and Mining Journal, 1933b: Reports placer-mining activity at Rabbit Hole; value of gravel and black sands mined; width and depth of placer gravel; type of false bedrock; mining techniques.

————1942: Reports presence of cassiterite in placers at Rabbit Hole, Placerites, and Sawtooth districts; notes that tin content is too low for profitable mining.

Lincoln, 1923: History; extent of placers (Rabbit Hole district); states the year that placers were first worked in Rosebud district.

Mining Journal, 1939b: Reports living arrangements of placer miners and number of miners in district.

————1940a: Reports amount of gold recovered daily by Rio Seco Mining Co.

Mining World, 1910: Reports placer-mining developments; average value of gravels per yard.

U.S. Bureau of Mines, 1936–40: Describes placer-mining operations in Rosebud district; Minerals Yearbook for 1936 states that Janke claims were in ancient lakebed gravels.

Vanderburg, 1936a: Rabbit Hole—early placer-mining activity and production (1916); depth of placer gravel; distribution of gold above and below false bedrock; size and fineness of gold; placer-mining activity and operations during the period 1932–35; amount of gold recovered by different operators. Rosebud—brief summary of placer-mining history and activity; depth of placer gravels.

————1936b: Early placer-mining history and production; names placer gulches; depth of gold-bearing gravel; fineness of gold; size of nuggets; distribution of gold; placer-mining operations.

91. SEVEN TROUGHS DISTRICT

Location: East flank of the Seven Troughs Mountains, T. 30 N., Rs. 28 and 29 E.

Topographic map: Lovelock 2-degree sheet, Army Map Service.

Geologic map: Tatlock, 1969, Preliminary geologic map of Pershing County, Nevada, scale 1:200,000.

Access: From Lovelock, 26 miles west on State Highway 48 to Vernon. Placers are accessible by dirt roads leading along flank of the mountains.

Extent: No descriptions of placers in the Seven Troughs district have been found, although placer gold was recovered periodically from 1913 to 1948. The placers are in the gulches and along the flanks of the range near many small gold and silver lode mines.

Production history: Except for the amount of placer gold produced, no information has been found describing placer mining in this district.

Source: The placer gold was derived from the gold veins in the area that contain free gold. Ransome (1909b, p. 22) states, "The valuable constituent of the lodes is native gold containing a considerable proportion of silver, and consequently of a rather pale color. In most of the rich ore the gold is visible either as clusters of small irregular particles or as coarse crystalline aggregates * * *. Loose nugget-like masses up to an ounce in weight have been found in soft crushed vein matter in the Reagan lease." Erosion of such veins would account for the origin of the placers in the area.

Literature:

Ransome, 1909b: Describes lode mines.

92. TRINITY DISTRICT

Location: On the east flank of the Trinity Range, between Black Rock Canyon and Trinity Canyon, Tps. 28 and 29 N., R. 31 E.

Topographic maps: Oreana 15-minute quadrangle; Lovelock 2-degree sheet, Army Map Service.

Geologic map: Tatlock, 1969, Preliminary geologic map of Pershing County, Nevada, scale 1:200,000.

Access: From Lovelock, 10 miles north on light-duty road to Trinity Canyon.

Extent: Small amounts of placer gold have been recovered from unlocated deposits in the Trinity district. The deposits probably occur in Trinity Canyon and nearby gulches, which drain areas where free gold is known to occur in lodes.

Production history: Placer gold was recovered intermittently between 1939 and 1963. Most of the work was apparently done by snipers using small-scale hand methods.

Source: The placer gold was probably eroded from silver-gold ores in which the gold is in the native state or is associated with iron oxides, as at the Evening Star mine (sec. 3, T. 28 N., R. 31 E.) and other lode mines in the vicinity.

Literature:

U.S. Bureau of Mines, 1957: States that placer mining was done with small-scale methods by miscellaneous prospectors and snipers.

Vanderburg, 1936b: Describes ores at Evening Star mine.

93. IMLAY (HUMBOLDT) DISTRICT

Location: West flank of the north end of the Humboldt Range between Prince Royal and Eldorado Canyons, Tps. 31 and 32 N., Rs. 33 and 34 E.

Topographic map: Imlay 15-minute quadrangle.

Geologic map: Silberling and Wallace, 1967, Geologic map of the Imlay quadrangle, Pershing County, Nevada, scale 1:62,500.

Access: From Lovelock, 37 miles north on Interstate 80 to Imlay. Numerous dirt roads lead from the highway east to the mining areas.

Extent: Placers occur in Imlay Canyon, Antelope Canyon, and probably in other canyons at the north end of the Humboldt Range. The placers in Imlay Canyon are near the Imlay mine (secs. 30 and 31, T. 32 N., R. 34 E.).

Production history: Substantial amounts of placer gold have been credited to the Imlay or Humboldt district between 1913 and 1951. For a few years between 1938 and 1949, however, placer gold actually recovered from gravels in Willow Creek (see 98, p. 83–84) was credited to the Imlay district. Placer-mining activity in the Imlay district was apparently restricted to small-scale drywashing of the gravels. The Willow Creek placers are richer than the Imlay Canyon placers, and it is probable that a considerable part of the placer production credited to the Imlay district actually originated in the Willow Creek district. I have changed the district data for those years where exact production can be credited to the proper district, but I estimate that at least 500 ounces credited to the Imlay district was produced from Willow Creek placers.

Source: The probable source of the placer gold in the Imlay district is gold-bearing veins that occur in Triassic sedimentary rocks in the region. The Imlay vein, the probable source of placer gold in Imlay Canyon, is composed of hard white quartz containing silver and gold.

Literature:

U.S. Bureau of Mines, 1930–31: Locates placers in Imlay and Antelope Canyons.

———1938–49: Describes placer-mining activity in Willow Creek under Imlay district.

94. UNIONVILLE DISTRICT

Location: East slope of the Humboldt Range, T. 30 N., R. 34 E.

Topographic map: Unionville 15-minute quadrangle.

Geologic map: Wallace, Tatlock, Silberling, and Irwin, 1969, Geologic map of the Unionville quadrangle, Pershing County, Nevada, scale 1:62,500.

Access: From Lovelock, 13 miles north on Interstate 80 to Oreana and junction with State Highway 50. From there, it is 16 miles east across the Humboldt Range and 12 miles north to Unionville on State Highway 50.

Extent: Placers are found in gravels in Buena Vista Canyon, above the abandoned townsite of Unionville, and in Congress Canyon, a tributary to Buena Vista Canyon from the north. The exact locations of the deposits in the canyons are unknown. A placer discovered in 1931 in Buena Vista Canyon consists of a 2- to 3-foot-thick pay streak in gravels 6–10 feet deep to bedrock. A placer worked in 1936 in Congress Canyon occurs in gravels 12 feet thick.

Production history: Recorded production for the Unionville district placers dated from 1940 to 1947; yet descriptions of the district indicate that some placers were worked as early as 1875, and others were actively worked between 1931 and 1936. Production during 1932 from placers in Buena Vista Canyon was said to be as high as $10 per day. Total production is probably twice that of recorded production of 46 ounces during the 20th century.

Source: The gold was probably derived from gold-bearing quartz fissure veins such as those worked at the Marigold mines (sec. 28, T. 30 N, R. 34 E.) in upper Buena Vista Canyon.

Literature:

Cameron, 1939: Source of gold; describes lode mines.

Raymond, 1877: Notes small-scale placer mining in Congress Canyon.

Vanderburg, 1936a: History of placer mining; placer-mining activity during the period 1931–35; depth of gravel worked in 1931; production per man per day in 1932.

————1936b: Placer mining in 1936; location; depth of gravel.

95. ROCHESTER DISTRICT

Location: West flank of the Humboldt Range, Tps. 28 and 29 N., Rs. 33 and 34 E.

Topographic maps: Unionville 15-minute quadrangle; Rochester mining district, special map, scale 1:24,000.

Geologic map: Wallace, Tatlock, Silberling, and Irwin, 1969, Geologic map of the Unionville quadrangle, Pershing County, Nevada, scale 1:62,500.

Access: From Lovelock, 13 miles north on Interstate 80 to Oreana and junction with State Highway 50; from there, it is about 8 miles east to placer areas north and south of Rochester.

Extent: Placers were worked in Sacramento, Limerick, Rochester, and Weaver Canyons on the west flank of the Humboldt Range. The placers were discovered in the early 1860's about the same time as the lode mines and worked on a small scale. Early placer activity was apparently concentrated in Rochester Canyon, but the placer workings are now covered by mill tailings from the lode mines. The gravels in Rochester Canyon below lower Rochester (secs. 13 and 14, T. 28 N., R. 33 E.) are 50 to several hundred feet thick and contain gold throughout the deposit. Gold is also reported in Tertiary gravels (sec. 23, T. 28 N., R. 33 E.) northwest of Packard Flat in the West Humboldt Range.

Placers in Limerick Canyon are the most productive in the district. Early placer miners are said to have worked the gravels in the canyon, but little information has been found about production from this location. Near the head of the canyon, west of Spring Valley Pass, an alluvial basin about 1½ miles wide and 1 mile long (secs. 4, 5, 8, 9, T. 28 N., R. 34 E.) known as Limerick Basin was the center for most placer-mining activity in the district in the 20th century. The gold is mainly concentrated on bedrock or in pay streaks just above bedrock; at the west end of the basin, however, gold was reported throughout the gravel thickness of 2–38 feet. Some gravel has yielded as much as $12 to $35 per cubic yard.

The placers in Sacramento Canyon (southern part of T. 29 N., Rs. 33 and 34 E.) in the north end of the Rochester district were discovered in 1912 and worked on a small scale.

Small placers were worked in Weaver Canyon (sec. 19, 20, 21, T. 28 N., R. 34 E.) in the south end of the Rochester district.

Production history: The production for the placers on the west side of the Humboldt Range (Rochester district, proper) is difficult to determine, because data for both Spring Valley and Rochester districts were grouped for most years. Early production has been estimated between 4,500 and 50,000 ounces; production during the 20th century is about 3,000 ounces.

Most of the placer mining in the Rochester district was on a small scale with drywashers, but a few small placer operations were mechanized. One operation in 1936 and 1937 at the Rhyolite placers in Limerick Canyon used ⅝-cubic-yard shovels and 2½-cubic-yard trucks to dig and transport the gravels to a stationary washing plant. Most of the work, however, was done by digging shafts to the pay streak and using drywashers to concentrate the fairly coarse gold.

Source: The Rochester district is primarily a silver mining district, but a few small lode-gold mines are found in the area. The gold is associated with quartz and tourmaline in veins that appear unrelated to the silver veins. In at least one place, placer gravels were traced to the lode source at the Hagan lode in Limerick Basin. The gold vein there contains quartz, microcline and tourmaline and cuts a quartz keratophyre phase of the Limerick Greenstone (Early Triassic). The gold veins are genetically associated with leucogranite (Early Triassic).

Literature:

Bergendahl, 1964: Placer-production estimate for Rochester district.

Gardner and Allsman, 1938: Depth of pay gravel and overburden; placer-mining techniques and operations at Rhyolite placer, Limerick Canyon.

Knopf, 1924: Brief description of history and production of placers in Limerick Basin; depth of gravel; average value; methods of mining; source.

Mining Journal, 1931: Reports sale and lease of placer ground in Rochester Canyon; price paid for land to Southern Pacific Company; thickness of gravel; average value per yard; purchase price of placer.

Southern Pacific Company, 1964: Locates placer gravels in the lower parts of Limerick and Rochester Canyons; indicates limited promise for future development.

Schrader, 1915: Describes placers in Rochester, Weaver, and Limerick Canyons. Details of depth of gravel, value, size, and production from Limerick Canyon.

Vanderburg, 1936a: Extent of placers; names of placer gulches; distribution of gold and thickness of gravels in Limerick Canyon; compares lithology of gravels in Limerick Basin with that of American Canyon.

————1936b: Placer-mining operations in 1930's in Limerick Canyon; depth of gravels worked in Limerick Basin; size and fineness of gold recovered.

96. SPRING VALLEY DISTRICT

Location: East flank of the Humboldt Range, Tps. 28 and 29 N., Rs. 34 and 35 E.

Topographic map: Unionville 15-minute quadrangle.

Geologic map: Wallace, Tatlock, Silberling, and Irwin, 1969, Geologic map of the Unionville quadrangle, Pershing County, Nevada, scale 1:62,500.

Access: From Lovelock, 13 miles north on Interstate 80 to Oreana and junction with State Highway 50. From there, it is about 16 miles east across the Humboldt Range to Fitting in lower Spring Valley Canyon. Placers are near Fitting and south along the flanks of the range.

Extent: Placers were worked in Spring Valley Canyon, Dry Gulch, and

American, South American, and Troy Canyons on the east side of the Humboldt Range. The Spring Valley district was formerly called the Indian Silver mining district, and it is under this district name that the first discovery of placer gold was recorded. Placers were discovered in Spring Valley Canyon in 1875, when the Eagle mine, now known as the Bonanza King mine, was one of the most active in the area. From the beginning, the placers were highly productive. Descriptions of placer-mining activity during the period 1875–76 state that in places 2 ounces of gold dust per day per man was recovered.

Spring Valley Creek heads near the crest of the Humboldt Range and flows east about 4 miles to Buena Vista Valley. The creek flows through Spring Valley, an alluvial basin, for about 1½ miles near the crest of the range. Most placer-mining activity was concentrated in the steep part of Spring Valley Creek, east of Spring Valley Basin (secs. 35 and 36, T. 29 N., R. 34 E., and sec. 31, T. 29 N., R. 35 E.). The gravels in the lower part of the canyon are 20–30 feet thick and contain gold in gravel horizons underlain by clay. The gold recovered from these gravels was coarse, and nuggets worth $3 to $5 were recovered in 1911. The fineness of the gold recovered in one placer operation ranged from 696 to 730.

Placers have been worked on a small scale in Dry Gulch, 1 mile south of Spring Valley Canyon (secs. 1 and 2, T. 28 N., R. 34 E.) since at least 1882.

American Canyon heads in the alluvial basin east of the crest of the Humboldt Range at Sage Hen Springs and flows about 4 miles east to Buena Vista Valley. Placers have been worked intensively from the edge of the range upstream for 2 miles (secs. 17 and 18, T. 28 N., R. 35 E.; sec. 13, T. 28 N., R. 34 E.). Most of the placer mining was done in the 1880's and 1890's by Chinese miners who dug shafts 40–100 feet into the gravels. The gold, both fine and coarse in size, was concentrated in a pay streak underlain by clay at an average depth of 60 feet but at shallower depth at the upper end of the placer than at the edge of the range. Little or no gold was found below the clay or on true bedrock. A continuation of the deposit was reportedly found in gravels overlain by lava near the edge of the range (sec. 17, T. 28 N., R. 35 E.). Schrader (1915, p. 368–370) describes placers overlain by basalts at the 4,700-foot elevation in Walker Gulch, half mile north of American Gulch (N½ sec. 17, T. 28 N., R. 35 E.). At the time of his visit, the deposit was being prospected along shafts 60 and 200 feet deep. The gold is described as particles generally valued at about one-fifth of a cent but reaching a maximum of 1½ cents, and the gravels were estimated to average (at that time) 75¢ to $1.00 per cubic yard.

South American Canyon is less than 1 mile south of American Canyon and joins American Canyon at the edge of the range. Placers in South American Canyon are found in a bowl-shaped alluvial area (probably in the S½ sec. 13, T. 28 N., R. 34 E.). Shafts in the sub-angular gravels were dug to depths of 15 feet.

Troy Canyon is 2 miles south of South American Canyon. Placers in this canyon were not so extensive nor productive as those in the northern canyons, and most of the work was apparently concentrated in gravels near the edge of range (secs. 31 and 32, T. 28 N., R. 35 E.).

Production history: The placers in the Spring Valley district are said to be the most productive in Nevada. Placer production estimated to be as much as $10 million, largely from American and Spring Valley Canyons, is attributed to early work by Chinese placer miners. Certainly, every description of the placer area credits the large amount of work done by these early miners, but production records for Humboldt County (before 1919, Pershing County was part of Humboldt County) do not reveal large amounts of gold. It is generally supposed that the Chinese miners never revealed the amount of placer gold recovered from the area, and, indeed, shipped large amounts of their proceeds out of the country.

The Chinese miners, who numbered in the hundreds, dug numerous shafts and systematically drifted along pay streaks and bedrock to recover the placer gold. The ground in American Canyon was worked by lessees who held blocks of land 20 feet square. Each block reportedly yielded $1,500 to $3,000 in gold.

Spring Valley Canyon has been the scene of most placer-mining activity in the district during the 20th century. Most of the work done over the years was small-scale drywashing, but two large-scale operations worked the lower part of the canyon. The first dredge in Spring Valley Canyon, also the first in Nevada, was the wooden dredge operated by the Federal Placer Mines Co. from 1911 to 1914. The dredge worked stream and bench gravels with an average recoverable value in gold of 31.6 cents per cubic yard. During the period 1947–49, the Spring Valley Gold Dredging Co., and then the Southwest Dredging Co., operated dryland dredges, dragline excavators, and a dryland washing plant in lower Spring Valley Canyon. The dredge tailings from this operation can still be seen at the mouth of the canyon. Both operations were successful, the operation in 1949 yielding about 2,000 ounces in placer gold.

Production figures for the 20th century from Dry Gulch and American, South American, and Troy Canyons are not available, but production has been small compared with that from Spring Valley Canyon.

Source: The only important lode mine in the Spring Valley district is the Bonanza King mine (formerly called the Eagle mine) located half a mile south of Spring Valley Canyon (NE¼ sec. 1, T. 28 N., R. 34 E.). The ore there contains gold, galena, pyrite, sphalerite, and tetrahedrite. Assuredly, erosion of parts of the Bonanza King vein must have contributed some gold to the placers in Spring Valley Canyon, but this same vein could not have been the source of the gold in the placers to the south.

Gold prospects occur along several small quartz-tourmaline veins which cut Rochester rhyolite at Gold Mountain, south of Spring Valley (center of sec. 3, T. 28 N., R. 34 E.). Similar quartz veins containing native gold were observed near the crest of the range (R. E. Wallace and D. B. Tatlock, written commun., 1971). These veins are related to swarms of rhyolite porphyry dikes but are in Limerick greenstone or Rochester rhyolite host rock.

It has been suggested by geologists who studied the area in the past that the gravels in American Canyon represent an ancient drainage that at one time crossed the Humboldt Range and connected with gravels in Limerick Basin and Canyon. Granite pebbles were reported in the American Canyon placers, although there is no granite exposed in that area. These pebbles may be from the leucogranite exposed in the Limerick Basin area, but this aspect of the geology of the Humboldt Range has not been studied in detail.

Literature:

Bergendahl, 1964: Placer-production estimate for Spring Valley district.

Burchard, 1883: Production from Spring Valley and Dry Gulches.

————1884: Production.

Lincoln, 1923: Gives dates of working placers in American Canyon; distribution of gold in gravels; dredge operations in Spring Valley in 1911.

Locke, 1913: Describes placers in American Canyon; production; early mining history; extent of placer; depth of pay streak; distribution of gold; notes placers found underlying lava.

Murbarger, 1958: Describes abandoned placer camp in American Canyon. Details how Chinese miners worked placer gravels by constructing shafts and drifts to false bedrock.

Paher, 1970: Brief history of early mining by Chinese miners; brief history of dredge mining; photograph of Federal dredge at work during the period 1910–14 included.

Ransome, 1909b: Describes extent of placer mining in American Canyon; depths of pay streaks; minerals associated with gold in placers.

Raymond, 1877: Notes placer production from gulch near Eagle mine.

Schrader, 1915: Describes placers in American Canyon, Walker Gulch, and Spring Valley and South American Canyons. Details of placer occurrence in gravels buried by basalt in Walker Gulch.

Southern Pacific Company, 1964: Located placer gravels at the mouth of Troy Canyon.

Vanderburg, 1936a: History; extent of placer workings; production; placer-mining operations; distribution of gold in gravels; size and fineness of placer gold; placer mining in 1936.

————1936b: History of placer mining in Spring Valley; placer-mining operations in 1930's in Dry Gulch and American and Spring Valley Canyons.

Walker, 1911: Describes dredge used in American Canyon; depth and average value of gravel; size of gold recovered; problems encountered in dredging.

Wallace and Tatlock, 1962: Outlines geologic setting for lode gold occurrence; notes situations for placer gold concentration.

Whitehill, 1877: Placer-mining activity in 1875 and 1876; locates deposits in relation to Eagle mine in Indian district (now Bonanza King mine in Spring Valley district); production per day per man; number of men working placers.

97. SIERRA (CHAFEY, DUN GLEN) DISTRICT

Location: Northern part of the East Range. Tps. 31–33 N., Rs. 36 and 37 E.

Topographic maps: Dun Glen and Rose Creek 15-minute quadrangles.

Geologic map: Ferguson, Muller, and Roberts, 1951b, Geologic map of the Winnemucca quadrangle, Nevada, scale 1:125,000.

Access: From Lovelock, 46 miles north on Interstate 80 to Mill City; from there, dirt roads lead northeast 8 miles to Dun Glen flat and placers along the west flank of the East Range.

Extent: Extensive placer deposits have been worked in Auburn and Wright Canyons (unlocated) and Barber Canyon, and Rockhill Canyon on the west flank of the East Range. Less extensive placers have been worked in Spaulding Canyon on the east flank of the range and in Dun Glen Canyon at the north end of the district. Willow Creek, lies between Spaulding Canyon and Rockhill Canyon but is considered to be a different district, more because of mining history than because of location.

The placers were discovered in the 1860's, and most of the mining was done during the period 1870–90 in Auburn and Barber Canyons, and during the period 1880–90 in Rockhill Canyon by Chinese miners. Little is known of the depth and value of the gravels mined at that time.

In the 1930's most of the placer-mining activity was concentrated in Dun Glen, Barber, and Spaulding Canyons. The gravels in these canyons are deep, ranging from 18 to 40 feet in Dun Glen Canyon and averaging 30 feet in Barber Canyon. The gold is generally found concentrated on bedrock and in some benches on the canyon sides.

Production history: The placers in the Sierra district are among the most productive in the State, the production being estimated at $4 million before 1900. This estimate represents the amount of gold thought to have been recovered by Chinese miners, who between 1870 and 1890 reportedly recovered $2 million from Auburn and Barber Canyons and between 1880 and 1895 recovered $2 million or more from Rockhill Canyon. As for many mining districts that were large producers before accurate records of mining activity were kept, there is some doubt that the actual production was as high as the estimated.

Placer mining during the 20th century was small scale and intermittent. So far as I know, no large-scale operations were successful, although a dryland dredge worked a short time in Dun Glen Canyon in 1931 and bulldozers and carryalls were used in Spaulding Gulch in 1940. Most of the placer gold was recovered by small-scale methods, such as sluicing, hydraulicking, and drywashing after drifting or stripping to the richer gravels near bedrock.

Source: The lode mines in the district are quartz veins carrying gold, silver, and sulfide minerals. Most are in the northern part of the district near the headwaters of Dun Glen and Barber Canyons (T. 33 N., Rs. 36 and 37 E.). Erosion of these veins, which locally contain high concentrations of gold, is the most likely source of the placer gold in the canyons. The veins appear to be post-Triassic and pre- or early Tertiary. Similar, but less conspicuous, veins probably supplied the placer gold in canyons south of the main lode-mining area.

Literature:

Ferguson and others, 1951: Briefly describes lode and placer deposits.

Lincoln, 1923: Early production of Chinese placer miners.

Mining Journal, 1940b: Reports beginning of production from placer operations in Spaulding Gulch; bulldozers and carryalls are used to move 2,000 yards of gravel per shift.

Vanderburg, 1936a: History of placer mining; names placer gulches, estimate of early production; placer-mining activity in Dun Glen, Barber, and Spaulding Canyons during the period 1931–34; depth and value of gravels in these canyons.

————1936b: Repeats placer description of earlier paper. Describes lode mines in district.

98. WILLOW CREEK DISTRICT

Location: West flank of the East Range, Tps. 31 and 32 N., R. 36 E.

Topographic map: Dun Glen 15-minute quadrangle.

Geologic map: Ferguson, Muller, and Roberts, 1951b, Geology of the Winnemucca quadrangle, Nevada, scale 1:125,000.

Access: From Lovelock, 46 miles north on Interstate 80 to Mill City; from there, dirt roads lead southeast about 12 miles to placers near the head-waters of Willow Creek.

Extent: Placers have been worked near the headwaters of Willow Creek, west of the crest of the East Range (secs. 2 and 11, T. 31 N., R. 36 E.). Some placers may have been worked farther downstream where the creek trends east-west (secs. 32–35, T. 32 N., R. 36 E.). The gold is apparently concentrated in channel fill of varying widths near bedrock and is overlain by alluvium. In the SE¼ sec. 11, in the upper part of Willow Creek where the three headward forks join to form the main creek, the overburden is from 15 to 30 feet thick.

Production history: Most placer activity took place between 1938 and 1964. Activity was concentrated at the Wadley placer mine (S½ sec. 11, T. 31 N., R. 36 E.) and the Thacker placer mine (N½ sec. 2, T. 31 N., R. 36 E.). Small earth-moving equipment was used to deliver the gravels to a central washing plant. In 1949 and 1950, Wallace Calder mined gravels containing $1.85 and 24¢ in gold per cubic yard at the Wadley mine. Dragline operations were carried on during the period 1959–60 in the north-south-trending part of the creek between the Thacker and Wadley placer mines.

Actual production from the Willow Creek placers is substantially higher than indicated by the recorded production of 2,823 ounces. For many years, placer production was included with that from the Imlay district and probably also with that from the Sierra district.

Source: Gold-bearing calcite and quartz veins exposed near the headwaters of Willow Creek are, at least in part, the source of the placer gold in Willow Creek. Both the Wadley and Thacker mines are in gravels mapped as older alluvium of Quaternary age that form small basins in the upper part of Willow Creek. These gravels represent remnants of gravel deposition from an earlier erosion cycle, and the gold found under the thick overburden was probably deposited at the earliest stage of the cycle.

Literature:

Ferguson and others, 1951: Notes mining activity in Willow Creek; source of placer gold.

Mining Journal, 1939a: Reports plans to use small dragline at Willow Creek.

Southern Pacific Company, 1964: Locates placers in Willow Creek; depth of overburden; states that the area warrants further investigation.

U.S. Bureau of Mines, 1938–63: Placer-mining activity at Willow Creek; some years give amount of gravel treated and ounces of gold and silver recovered; names placer claims.

OTHER DISTRICTS

99. GOLDBANKS DISTRICT

The eastern edge of the Goldbanks Hills at the north end of Pleasant Valley (secs. 20 and 21, T. 30 N., R. 39 E.) contain gold-silver ores in quartz veins within rhyolite that were prospected on a small scale during the period 1907–8. Gravels at the foot of the hill in which the gold lodes are situated probably were the source for the small amount of placer gold recovered from the area in 1949.

Literature:

Dreyer, 1940.

100. KENNEDY DISTRICT

The placer gold credited to the Kennedy district probably was recovered from gravels in Kennedy Canyon or its tributaries (T. 28 N., R. 38 E.) on the east side of the East Range. The placer gold was probably derived from gold-silver ores in veins in the igneous rocks of the area that have been mined near the head of the canyon, particularly at the Gold Note mine during the 1890's.

Literature:

Ransome, 1909b.

Vanderburg, 1936b.

101. MILL CITY (CENTRAL) DISTRICT

The Mill City district is in the Eugene Mountains west of the Humboldt River in Pershing and Humboldt Counties. The area contains productive tungsten and lead mines and a few small gold prospects. The location and source of the placer gold credited to the district from both Pershing and Humboldt Counties are unknown. Most of the gold known to have been mined in the district apparently occurs in the northern part of the range in the area known as the Central district, Humboldt County.

Literature:

Willden, 1964.

Vanderburg, 1936b.

102. STAR DISTRICT

Placers were discovered in Star Creek, northeast of Star Peak on the eastern flank of the north end of the Humboldt Range in 1868; the most valuable deposits were found in creek gravels at the range front (sec. 24, T. 31 N., R. 34 E.), but some gold was found in gravel banks near the silver mines 2½ miles above (west) of the mouth of the canyon. The gold in the gravels in the lower part of the creek was pure, coarse, and rough. The Star district is noted for silver deposits upstream from the placers, and early miners searched for gold lodes that must have been the source, apparently without success. No production figures were found for the placers reportedly worked between 1868 and 1871.

Literature:

Raymond, 1870, p. 192–193.

———1873, p. 209.

103. STAGGS DISTRICT

The Staggs district is in the northern Black Mountains, also known as the Bluewing Mountains, in western Pershing County. A small amount of placer gold was recovered during the period 1940–41 from a placer deposit at the northwest end of the mountains (sec. 12, T. 29 N., R. 26 E.). The placer extends north along the major drainage from the Bluewing Mountains.

STOREY AND ORMSBY COUNTIES

104. COMSTOCK DISTRICT (STOREY COUNTY)

Location: In the Virginia Range, northwest of the Carson River, T. 17 N., R. 21 E.

Topographic map: Virginia City 15-minute quadrangle.

Geologic map: Thompson, 1956, Geologic map and sections of the Virginia City quadrangle, Nevada, scale 1:62,500.

Access: From Reno, about 9 miles southeast on U.S. Highway 395 to junction with State Highway 17; from there, about 12 miles southeast on State Highway 17 to Virginia City and Comstock district.

Extent: Placer gold occurs in Gold and Six Mile Canyons in the Comstock district. The placer gold credited to this district may have been recovered from gravels in these drainages or from old tailings found throughout the mining area. No information has been found to indicate the location or character of the deposits, mined since 1900.

Production history: Placer gold was recovered from the Comstock district during the initial prospecting stage (about 1857–59) in this famous lode-mining area. The discovery of the Ophir mine (sec. 29) is attributed to placer miners working the gravels of Six Mile Canyon to its head. Decomposed ores of the Sierra Nevada mine (sec. 20), near Seven Mile Canyon, tributary to Six Mile Canyon, were mined by placer methods during the 1860's. Placering during the 20th century is reported for the period 1934–47, but, except for the year 1934, when 354 ounces of gold was recovered by two operators, yearly production was very small.

Source: The source of the placer gold is certainly the ores of the Comstock lode; it is not known whether the gold was truly erosional material from the lode or from particles contained in old tailings.

Literature:

Bonham, 1969: Summarizes earlier work on geology of the Comstock lode; dates mineralization of lode.

De Quille, 1891: Describes placer mining in Six Mile Canyon; discovery of other lodes.

Paher, 1970: History of early placer mining in Six Mile Canyon; production per day by O'Reiley and McLaughlin in gravels in Ophir Lode.

Stoddard and Carpenter, 1950: Describes ore deposits of Comstock lode.

Thompson, 1956: Notes extent of placers; describes lode mines.

White, 1871: Describes gold-bearing gravel at Sierra Nevada mine.

105. CARSON RIVER DISTRICT (ORMSBY COUNTY)

Location: Along the Carson River, in the vicinity of Empire, T. 15 N., Rs. 20 and 21 E.

Topographic map: Dayton 15-minute quadrangle.

Geologic map: Moore, 1969, Geologic map of Lyon, Douglas, and Ormsby Counties, Nevada (pl. 1), scale 1:250,000.

Access: From Carson City, 3 miles northeast on State Highway 50 to vicinity of Carson River at Empire.

Extent: Gravels along the Carson River have been mined sporadically for placer gold that in part came from old mill tailings from treatment of the Comstock ores. The gold recovered during the 20th century apparently came from the part of the river in the vicinity of Empire (secs. 11–14, T. 15 N., R. 20 E.) and near Santiago Canyon (sec. 5, T. 15 N., R. 21 E.). In 1923, an old 60-foot-wide channel of the Carson River was reportedly discovered to contain coarse gold.

Production history: Production from the Carson River placers has been minor. The many attempts to recover gold from old mill tailings along the Carson River before 1900 met with failure, and placer mining during the 20th century has been sporadic.

Source: The probable source of the placer gold is the Comstock lode.

Literature:

Mining Review, 1923: Reports discovery of old river channel of Carson River on the old Mexican Mill property at Empire; channel gravels reportedly contain coarse gold; width of channel bed is 60 feet.

U.S. Geological Survey, 1911: Notes placer production from Carson River; states that gold is presumed to have been recovered from old tailings.

Vanderburg, 1936a: History of operations along Carson River to recover gold in mill tailings from Comstock lode.

WASHOE COUNTY

106. JUMBO (WEST COMSTOCK) DISTRICT

Location: West flank of the Virginia Range, T. 16 N., R. 20 E.

Topographic map: Virginia City 15-minute quadrangle.

Geologic map: Thompson, 1956, Geologic map of the Virginia City quadrangle, Nevada (pl. 3), scale 1:62,500.

Access: From Reno, 16 miles south on U.S. Highway 395 to junction with dirt road that leads east, up the flank of Virginia Range, about 6 miles to Jumbo mining district.

Extent: The Jumbo district is a small lode mining district on the opposite flank of the Virginia Range from the Comstock district. Most of the mines are near the crest of the range (secs. 34 and 35, T. 16 N., R. 20 E.). Small amounts of placer gold were recovered from placers in this district.

Production history: Most of the placer gold credited to the Jumbo district was recovered between 1937 and 1940, by small-scale methods.

Source: The placer gold in the Jumbo district was probably derived from the small oxidized veins that cut the Alta Formation (Miocene). Free gold is the only economically important metal in the veins.

Literature:

Bonham, 1969: Describes lode deposits; notes small-scale placer mining.

Mining and Engineering World, 1913b: Reports plans to develop lodes and placers; extensive placer gravels exposed near Pandora Comstock lode.

Thompson, 1956: Briefly describes types of ores found in Jumbo district.

U.S. Bureau of Mines, 1937–40: Gives placer production figures.

107. OLINGHOUSE (WHITE HORSE) DISTRICT

Location: East flank of the Pah Rah Range, T. 21 N., R. 23 E.

Topographic map: Wadsworth 15-minute quadrangle.

Geologic map: Bonham, 1969, Geologic map of Washoe and Storey Counties, scale 1:250,000.

Access: From Reno, 32 miles east on Interstate 80 to Wadsworth; from there, 3 miles north on State Highway 34 to dirt road in Olinghouse Canyon leading 6 miles west to mining area.

Extent: Gold placers are found in an alluvial basin about 1 mile long and ½ mile wide and in a tributary ravine north of Olinghouse Canyon and south of Green Hill in the main lode mining area of the district (secs. 20–29, T. 21 N., R. 23 E.). Gold-bearing placers also occur at the edge of the range near Frank Free Canyon east of Green Hill (sec. 27, T. 21 N., R. 23 E.) and Tiger Canyon north of Green Hill (sec. 21, T. 21 N., R. 23 E.).

The gravels in the alluvial basin and tributary ravine average about 20 feet deep. The gold is concentrated in the lowermost 5–6 feet of gravel above bedrock. Along the south and east margins of Green Hill, the placers are eluvial, whereas those in drainages from Green Hill are alluvial deposits transported 1 mile or more. Near the edge of the range at Frank Free Canyon, shafts and churn drills sampled gravels to a depth of 75 feet that assayed 8, 23, and 94 cents per cubic yard.

Production history: The placers in the Olinghouse district were extensively worked between 1860 and 1900 and were said to produce considerable gold, but no authentic records of placer gold production are known. The total placer and lode-gold production of the district before 1900 has been estimated at about $218,000, although some estimates indicate as much as $500,000. During the 20th century, the Olinghouse placers have been worked almost continuously, but mostly on a small scale. The rich gravels were reached by drifts concentrated in the center of the alluvial basin, where they apparently follow a channel to the south towards Olinghouse Canyon. In a flat part of the alluvial basin (approximately north edge of sec. 29), a small dragline dredge worked the gravels in 1965 by digging a square pit about 15 feet deep, then floating the dredge in the water-filled pit and back-filling with tailings. The operation was not successful because of difficulty in keeping water in the pit. Another operation dredged gravels during the period 1963–64 in the tributary ravine (east side of the road to the mining area, center sec. 29).

Source: Small gold-bearing quartz and calcite veins in the andesites and intrusive granodiorite porphyry (Miocene and Pliocene) at Green Hill are the source of the placer gold. The gravels in which the gold is found consist mostly of subangular andesite and basalt debris derived from the adjacent hillsides.

Literature:

Bonham, 1969: History; placer-mining activity during the period 1963–65; distribution of placers; describes details of lode mines.

Engineering and Mining Journal, 1897d: Reports production of $207 in gold from 12 tons of placer gravel in Olinghouse Canyon.

Hill, 1911: Estimate of lode and placer production; describes lode mines.

Overton, 1947: Describes lode deposits; source of placer gold; placer-mining activity in 1937.

Southern Pacific Company, 1964: Locates area for future exploration; value of gold per cubic yard in samples.

U.S. Bureau of Mines, 1963–64: Describes placer operations with dragline dredge; names placer claims.

Vanderburg, 1936a: Early placer-mining history; placer-mining operations in 1935; location of these operations; depth of placer gravel;
size of gravels; fineness and size of gold; source, methods of mining;
average value of gravel.

108. PEAVINE DISTRICT

Location: Northeast slope of Peavine Peak, T. 20 N., Rs. 18 and 19 E.

Topographic map: Reno 15-minute quadrangle.

Geologic map: Bonham, H., 1969, Geologic map of Washoe and Storey
Counties, scale 1:250,000.

Access: From Reno, 8 miles north on U.S. Highway 395 to dirt road leading southwest to flank of Peavine Peak.

Extent: Small placer deposits occur in gulches and ravines on the northeast slope of Peavine Peak in the area where quartz monzonite is exposed. Only sporadic mining has been done on these deposits during
this century, and it is impossible to locate all the placers. Before 1900,
the Nevada Industrial Placer (SW¼ sec. 16, T. 20 N., R. 19 E.) was
actively mined. This placer, located in a small ravine, has been worked
over an area 1,500 feet long and 2–3 feet wide.

Production history: The placers in the Peavine district were reportedly
worked between 1876 and the 1890's, but no authentic records of production have been found. Several thousand dollars of placer gold were
reported to have been recovered from the Nevada Industrial placer.
During the 20th century, placer mining has been very sporadic, and the
amount of gold recovered small. The largest amount of gold recovered
in 1 year was in 1957, when prospectors and snipers recovered 21 ounces
from stream deposits.

Source: The placer gold was derived from replacement deposits in quartz
monzonite formed during the late Miocene and Pliocene. These deposits
consist of narrow zones of magnetite and small stringers of pyrite that
probably contain gold; erosion of these deposits is said to supply sufficient gold to account for the placers.

Literature:

Bonham, 1969: Describes lode mines; dates age of mineralization.

Hill, 1915: Production estimate; locates Nevada Industrial placer property; length and width of placer area; source.

Overton, 1947: Describes lode deposits; describes type of lode deposit
which probably was the source of placers, locates placers.

Vanderburg, 1936a: Extent of placers; states that records of placer
mining exist for 1876–78 and for the 1890's.

White, 1871: Notes placer-mining activity when water was available
(Peavine district, p. 4).

109. GALENA (WASHOE) DISTRICT

Location: East flank of the Carson Range, west of Pleasant Valley, T. 17 N., R. 19 E.

Topographic map: Mount Rose 15-minute quadrangle.

Geologic map: Thompson and White, 1964, Geologic map and sections of the Mount Rose quadrangle, Washoe County, Nevada (pl. 1), scale 1:62,500.

Access: From Reno, about 14 miles south on U.S. Highway 395 to Pleasant Valley. Placers are in hills west of the highway.

Extent: Small placer deposits are located near the mouth of Galena Creek (sec. 12, T. 17 N., R. 19 E.) and along Steamboat Creek, near Little Washoe Lake (sec. 24).

Production history: The placer production intermittently credited to the Galena district has been small.

Source: The gold recovered from Galena Creek was apparently derived from old tailings, probably from the Union lead mine which contains gold and silver associated with more abundant sulfides. The origin of the gold recovered from Steamboat Creek is unknown.

Literature:

Bonham, 1969: Describes and locates lode mines.

Thompson and White, 1964: Describes ore deposits in Galena district.

110. LITTLE VALLEY DISTRICT

Location: East flank of the Carson Range in Little Valley, T. 16 N., R. 19 E.

Topographic maps: Carson City and Mount Rose 15-minute quadrangles.

Geologic map: Thompson, and White, 1964, Geologic map of the Mount Rose quadrangle, Washoe County, Nevada (pl. 1), scale 1:62,500.

Access: From Reno, 18 miles south on U.S. Highway 395 to the site of Franktown (4 miles south of Washoe City); jeep trails lead west from Franktown up the flank of the mountain to Little Valley.

Extent: Gold-bearing gravels that underlie Tertiary volcanic rocks were mined before 1900 in Little Valley. Most of the placers are in the southern part of Little Valley (SW¼ T. 16 N., R. 19 E., Carson City quadrangle). According to Reid (1908), the gold is found as well-rounded grains ranging in size from that of a mustard seed to coarse nuggets.

Production history: No reliable production figures have been found. Reid reports an estimated production of $100,000 in placer gold. Shallow cuts in one area in Little Valley are said to have yielded $60,000 in placer gold from a pocket in the gravel.

Source: The source of the gold is unknown.

Literature:

Bonham, 1969: Virtually repeats Vanderburg (1936a).

Reid, 1908: Describes extent of Tertiary river channel; placer-mining history; size of gold in gravels; estimate of early production.

Thompson and White, 1964: Approximately locates placers.

Vanderburg, 1936a: Placer-mining history; extent and significance of Tertiary gravels; estimates of production; placer-mining activity in 1935.

WHITE PINE COUNTY

111. BALD MOUNTAIN (JOY) DISTRICT

Location: Western slope of the southern end of the Ruby Range, T. 24 N., Rs. 56 and 57 E. (projected).

Topographic map: Gold Creek Ranch 15-minute quadrangle.

Geologic map: Rigby, 1960, Preliminary geologic map of the Bald Mountain area, Nevada (fig. 5), scale ≈ 1:125,000.

Access: From Ely, 59 miles west on U.S. Highway 50 to improved road leading north 35 miles through the Newark Valley. At the north end of the valley, dirt roads lead about 8 miles east and north to placer area at flanks of Bald Mountain.

Extent: Placer gold occurs in gravels of Water Canyon on the south slope of Big Bald Mountain. Apparently, most of the placers occur above the 7,000-foot elevation at the base of the mountain, although some reports indicate that some gold has been found for 6 miles along the Canyon. The gravels in the narrow canyon east of the 7,000-foot elevation are about 10 feet thick. The pay streak on bedrock was 14–18 inches thick. The gold reportedly was coarse, and nuggets ranging in value from $2.50 to $10 were found.

Production history: Although the Bald Mountain placers are mentioned in a number of reports, production was probably small. One report states that men working the area in 1933 recovered at least 1 ounce of gold per shift, but no production was recorded by the U.S. Bureau of Mines that year.

Source: The probable source of the placer gold is from gold veins in the quartz monzonite of Bald Mountain, yet the placer gold is said to be coarser than the gold in these veins.

Literature:

Blake, 1964: States that placer gold is coarser than remaining veins in probable source—quartz monzonite (p. 29).

Hill, 1916: Thickness of gold-bearing gravels; size of nuggets recovered.

Mining Review, 1933: Reports developments at lode and placer properties; names operators at different placers; gulches in which placers are located; production from one claim.

Vanderburg, 1936a: Early mining history; depth and extent of placer gravels; thickness of pay streak; size of nuggets found.

112. CHERRY CREEK DISTRICT

Location: Northern Egan Range, T. 23 N., R. 62 E.

Topographic map: Ely 2-degree sheet, Army Map Service.

Geologic map: Hose and Blake, 1970, Geologic map of White Pine County, Nevada, scale 1:250,000.

Access: From Ely, 44 miles north on U.S. Highway 93 to junction with State Highway 35; Cherry Creek is at the terminus of State Highway 35, 8 miles west of U.S. Highway 93. From Cherry Creek, Egan Canyon is reached by dirt roads leading 3 miles south.

Extent: Placer gold occurs in the gravels along Egan Canyon (center T. 23 N., R. 62 E.), but the exact location of the deposits is not known.

Production history: Placer gold was known in the canyon before 1916. A small production was recorded for 1932.

Source: The sources of the placer gold are veins in the vicinity of Egan Canyon. These veins are of two types—free gold in quartz veins and silver-gold base-metal veins.

Literature:

Hill, 1916: Notes presence of placer gold in Egan Canyon; does not describe occurrence.

113. OSCEOLA DISTRICT

Location: West flank of the central part of the Snake Range, south of Sacramento Pass, T. 14 N., Rs. 67 and 68 E.

Topographic map: Sacramento Pass 15-minute quadrangle.

Geologic map: Hose and Blake, 1970, Geologic map of White Pine County, Nevada, scale 1:250,000.

Access: From Ely, 34 miles south and east on U.S. Highway 50 to dirt road leading to Osceola, 4 miles east of the main highway.

Extent: Thick deposits of gold-bearing gravels are on the west slope of the Snake Range in the vicinity of Dry Gulch and Mary Ann Canyon (east half of T. 14 N., R. 67 E.). Less extensive placers are found in gravels of Weaver Creek and the Summit diggings (W½ T. 14 N., R. 68 E.) on the east flank of the Snake Range. The most productive placers of the Osceola district are concentrated in the two areas on the west flank. The placers in Dry Gulch and Grub Gulch (formerly Wet Gulch) in secs. 11 and 12, T. 14 N., R. 67 E., were first discovered in 1877. There the gravels range from a thin covering on quartzite bedrock to more than 200 feet thick. The gold is found distributed throughout the gravel thickness, but highest values are concentrated on bedrock; at the Hampton placer in the upper part of Dry Gulch, gold values ranged from 17¢ to $8.77 per cubic yard from surface to bedrock (Vanderburg, 1936a, p. 169).

The placers at Mary Ann Canyon are known as the Hogum placers. These deposits are in the alluvial fan at the mouth of Mary Ann Canyon, 3 miles south of Osceola (secs. 23, 24, 26, T. 14 N., R. 67 E.). These deposits were discovered some years after the placers in Dry Gulch (probably in 1879); the gold-bearing gravels occur in buried channels under the gravels of the alluvial fan but overlying cemented gravel layers which occur at different levels.

The placers at Summit diggings (sec. 8 or 9, T. 14 N., R. 68 E.) and Weaver Creek (sec. 10 or 15, T. 14 N., R. 68 E.) were not so extensive nor so profitable as those at Dry Gulch and Mary Ann Canyon.

Production history: Since the discovery of the Osceola placers in 1877, placer mining has continued in the district with different methods and intensity. In the decades following the discovery, hydraulic placer mining was successful at Dry Gulch. During the late 1930's, hydraulic mining at the Hampton placer in Dry Gulch that had been hydraulicked in the early days produced the highest yearly total of placer gold recorded for the district.

The placers in the Hogum area were usually mined by sinking shafts and drifting through the gravels to reach the channels containing the highest concentrations of gold. Although most of the placer gold recovered from the Osceola district was fine in size, several very large nuggets were recovered during the past century.

Source: The source of the placer gold is the lode deposits that occur in Cambrian quartzites filling regular fractures, or as sheeted zones or irregular shattered masses. The most important lodes occur upstream from the most productive placers—on the ridge west and south of Dry Gulch (secs. 12 and 13, T. 14 N., R. 67 E., and sec. 18, T. 14 N., R. 68 E.) and on the slopes of Mary Ann Canyon (secs. 25 and 26, T. 14 N., R. 67 E., and secs. 19 and 30, T. 14 N., R. 68 E.). Free gold is the only commercial metal in the ores.

Literature:

Burchard, 1884: States that a placer was discovered in 1883; size of large nugget; extent of placer ground; yield per day per man.

————1885: Placer-mining operations; developments by Osceola Gravel Mining Co.; value of gold in deep bars; notes hydraulic operations.

Engineering and Mining Journal, 1887: News note abstracts professional report by George Maynard; extent of placer ground in acres; equipment on property; thickness of gravel; production from hydraulic mine; value of gravel on bedrock; average value of gravel; size of large nuggets recovered; potential developments discussed.

————1891: Partial production; size of large nugget recovered (53 oz.).

————1892a: Size of nugget found on November 29, 1892; weight of gold in nugget (125 oz.); valued at $2,200.

————1892b: Reports recovery of 35-ounce gold nugget with some quartz attached; valued at $550.

Mining Review, 1910: Report of renewed large-scale operations at Osceola by Gold Bar Co.

Paher, 1970: States that placers were discovered in 1872; brief history of Osceola Placer Mining Co. operation; states that nugget valued at $6,000 was found in 1886; photograph of hydraulic mining.

Stuart, 1909: History of placer discovery; size of large nugget (24 lb.) recovered in 1878; production estimates; number of men working gravels; principal placer areas; distribution of gold in Mary Ann Canyon.

Vanderburg, 1936a: History; detailed descriptions of certain placer mines; average value of gravels; thickness of gravels; methods of working gravels; placer-mining operations during the period 1932–35; problems associated with placer mining.

Weeks, 1908: History; early production estimates; distribution of placers; thickness of gravels; size of gold; describes gold veins from which placers were derived; summarizes placer-mining activity from 1877 to 1907.

Whitehill, 1879: Date and location of placer discovery; size of large nugget recovered; placer-mining operations; number of claims located.

OTHER DISTRICTS

114. GRANITE DISTRICT

Placer gold is reported from the Granite district on the east side of the southern Egan Range (T. 19 N., Rs. 62 and 63 E.). No information is known about the placer occurrence; free gold reportedly occurs in the veins in the district.

Literature:

Hill, 1916.

Hose and others, 1972.

115. ROBINSON DISTRICT

Placer gold was credited to the Robinson district in 1909, but U. S. Bureau of Mines records state that the district of origin is unknown. The Robinson district, west of Ely in the Egan Range (T. 16 N., R. 62 E.), is noted for copper ores containing gold, silver, and other metals. The placer gold may have been mined in outlying districts, such as the Granite district located 20 miles north, and sold to buyers in the Robinson district.

GOLD PRODUCTION FROM PLACER DEPOSITS

Nevada ranks sixth in the United States and fifth in the western continental States in placer-gold production. The U.S. Bureau of Mines (1967, p. 15) cites 1,900,000 troy ounces of placer gold produced in Nevada from 1792 to 1964. I estimate a total production of 1,700,000 ounces of placer gold for the State from the first placer discovery to the present (table 1). The U.S. Bureau of Mines estimate includes some unauthenticated reports of very high placer gold production from some districts worked before 1900.

TABLE 1.—*Nevada placer gold production, in ounces*

Map locality (pl. 1)	County and placer districts	Estimated production, discovery to 1901	Recorded production data from U.S. Bur. Mines			Total recorded production 1902–68	Total estimated production	Reference source for estimated production
			1902–33	1934–42	1943–68			
	Churchill:							
1	Holy Cross	0	12	0	63	75	100	
2	Jessup	0	0	1	0	1	5	
3	Sand Springs	0	0	0	10	10	10	
	Clark:							
4	Eldorado Canyon	300	151	18	4	173	470	Vanderburg (1936a).
5	Gold Butte	0	0	1	0	1	10	
6	Las Vegas	0	43	0	0	43	43	
7	Searchlight	0	0	24	2	26	50	
8	Boulder Dam	0	0	1	0	1	1	
9	Bunkerville	0	0	0	0	0	0	
10	Muddy River	0	6	0	0	6	6	
	Douglas:							
11	Mount Siegel	5,000	910	23	3	936	6,000	Do.
12	Genoa	0	100	0	0	100	100	
13	Mountain House (Pine Nut).	0	1	3	0	4	4	
	Elko:							
14	Centennial	0	5	11	0	16	30	
15	Charleston	≈300	47	94	0	141	500	
16	Cope (Mountain City)	<200	0	0	0	0	<250	Production included with Van Duzer district.

No.								Reference
17	Gold Circle (Midas)	0	57	1	0	58	58	
18	Jarbidge	0	1	5	0	6	10	
19	Island Mountain	40,000	7	315	418	740	41,000	Murbarger (1957).
20	Tuscarora	35,000	776	67	0	843	36,000	Nolan (1936b).
21	Van Duzer	2,500	1,289	2,325	834	4,448	10,000	Vanderburg (1936a):
22	Alder	Unknown	0	0	0	0	Unknown	
23	Gold Basin	Unknown	0	0	0	0	Unknown	
	Esmeralda:							
24	Lida (Tule Canyon)	Unknown	211	1,377	90	1,678	3,000–5,000	
25	Sylvania	Unknown	93	197	8	298	500	
26	Desert	0	0	29	0	29	29	
27	Goldfield	0	115	4	4	123	123	
28	Gold Mountain (Tokop)	Unknown	2	0	0	2	2	
29	Hornsilver	0	0	0	16	16	16	
30	Klondyke	0	0	0	0	0	Unknown	
	Eureka:							
31	Lynn	0	5,684	3,696	563	9,943	10,000	
32	Eureka	0	0	2	0	2	2	
33	Maggie Creek	0	0	0	0	0	Unknown	
	Humboldt:							
34	Sawtooth	0	113	783	45	941	1,000	
35	Dutch Flat	3,750	43	20	244	307	4,500	Do.
36	Gold Run	Unknown	1,289	275	308	1,872	2,000	
37	Rebel Creek and National.	Unknown	7	13	0	20	Unknown	
38	Varyville (Leonard Creek).	0	39	129	0	168	200	
39	Awakening	0	12	1	0	13	25	
40	Dunnashee	0	0	0	6	6	10	

TABLE 1.—*Nevada placer gold production, in ounces*—Continued

Map locality (pl. 1)	County and placer districts	Estimated production, discovery to 1901	Recorded production data from U.S. Bur. Mines			Total recorded production 1902–68	Total estimated production	Reference sour fo estimated production
			1902–33	1934–42	1943–68			
	Humboldt—Continued							
41-------	Jackson Creek-------	0	0	4	0	4	4	
42-------	Kings River (Disaster)----	0	0	0	0	0	5	
43-------	Potosi-------	0	0	0	C	C	25	
44-------	Warm Springs-------	0	0	21	0	21	25	
45-------	Winnemucca-------	0	97	33	0	130	130	
	Lander:							
46-------	Battle Mountain-------	0	44,292	8,983	>52,730	106,005	156,000	
47-------	Bullion-------	0	70	10,258	0	10,328	11,000	
48-------	Hilltop-------	0	205	26	0	231	231	Do.
49-------	McCoy-------	0	4	37	0	41	41	
50-------	Birch Creek-------	0	0	0	0	0	0	
51-------	Iowa Canyon-------	0	0	0	0	0	0	
52-------	Kingston-------	0	0	3	0	3	3	
53-------	Reese River-------	0	3	0	0	3	3	Do.
54-------	Steiner Canyon-------	0	0	0	0	0	0	
	Lincoln:							
55-------	Eagle Valley-------	0	0	2	0	2	2	
56-------	Freiburg-------	0	0	13	0	13	13	
	Lyon:							
57-------	Buckskin-------	0	7	75	9	91	100	
58-------	Yerington-------	0	54	166	0	220	250	

No.	Locality	>26,500 Unknown	14,990	38,085	5,944	59,019	90,000–100,000 Unknown	Lord (1883).
59	Silver City	Unknown	0	0	0	0	Unknown	
60	Como	0	0	3	0	3	3	
61	Eldorado Canyon	0	2	1	0	3	3	
62	Pine Grove	0	4	0	0	4	4	
63	Talapoosa	0						
	Mineral:							
64	East Walker	0	0	235	0	235	235	
65	Hawthorne	0	145	51	8	204	204	
66	Rawhide	0	1,737	292	141	2,170	10,000	
67	Aurora	0	0	4	0	4	4	
68	Bell	0	0	2	0	2	2	
69	Candelaria	0	46	0	0	46	46	
70	Santa Fe	0	12	0	0	12	12	
71	Silver Star	0	21	0	0	21	21	
72	Telephone Canyon	0	0	0	0	0	5	
	Nye:							
73	Bullfrog	0	33	6	1	40	45	
74	Johnnie	0	81	249	91	421	500	
75	Cloverdale	0	22	52	1	75	100	
76	Ione (Union)	0	34	17	0	51	100	
77	Millett	0	23	0	0	23	23	
78	Manhattan	0	62,710	100,169	44,019	206,898	210,000	
79	Round Mountain	0	65,382	11,834	110,301	187,517	232,000	
80	Belmont	0	0	0	0	0	0	
81	Currant	0	1	0	0	1	1	
82	Eden Creek	0	0	7	0	7	7	
83	Ellendale	0	0	2	0	2	2	
84	Fairplay	0	0	0	(¹)	(¹)	≈10	

¹Quantity confidential.

TABLE 1.—*Nevada placer gold production, in ounces*—Continued

Map locality (pl. 1)	County and placer districts	Estimated production, discovery to 1901	Recorded production data from U.S. Bur. Mines			Total recorded production 1902-68	Total estimated production	Reference source for estimated production
			1902-33	1934-42	1943-68			
	Nye—Continued							
85	Lodi (Mammoth)	0	0	33	0	33	33	
86	Longstreet	0	0	0	0	0	5	
87	Tonopah	0	0	2	0	2	2	Includes Esmeralda County production.
	Pershing:							
88	Antelope	0	0	625	19	644	700	
89	Placerites	1,500	153	267	134	554	2,500	Vanderburg (1936a)
90	Rosebud and Rabbithole	0	666	6,930	317	7,913	8,000	
91	Seven Troughs	0	143	34	26	203	250	
92	Trinity	0	0	84	13	97	100	
93	Imlay (Humboldt)	0	210	<756	<579	<1,545	1,000	
94	Unionville	100	0	39	7	46	200	
95, 96	Rochester and Spring Valley.	500,000	6,655	1,617	2,710	10,982	511,000	Ransome (1909b);
97	Sierra	200,000	262	205	67	534	200,500	Lincoln (1923).
98	Willow Creek	0	0	>1,155	>1,668	>2,823	4,000	
99	Goldbanks	0	0	0	5	5	5	
100	Kennedy	0	34	0	0	34	34	

101 Mill City	0	0	8	116	124	124
102 Star	50	0	0	0	0	50
103 Staggs	0	0	7	0	7	7
Storey and Ornsby:						
104 Comstock	Unknown	0	538	45	583	1,000
105 Carson River	Unknown	43	0	14	57	<100
Washoe:						
106 Jumbo	0	6	34	0	40	50
107 Olinghouse	Unknown	254	248	334	836	≈2,000
108 Peavine	≈150	9	2	21	32	≈200
109 Galena	0	14	5	0	19	<25
110 Little Valley	5,000	0	0	0	0	5,000 Reid (1908).
White Pine:						
111 Bald Mountain	0	0	4	0	4	<10
112 Cherry Creek	0	4	0	0	4	<10
113 Osceola	85,500	3,253	3,125	215	6,593	95,000 Weeks (1908).
114 Granite	0	27	0	0	27	27
115 Robinson (Ely)	0	9	0	0	9	9
Total	905,850	212,730	195,763	222,153	630,646	≈1,700,000
Undistributed to districts	------	5,608	3,013	99,576	108,197	------
State totals	905,850	218,338	198,776	321,729	738,843	≈1,700,000

The most productive placer districts in Nevada are the Battle Mountain district, Lander County; Silver City district, Lyon County; Manhattan and Round Mountain districts, Nye County; Spring Valley and Sierra districts, Pershing County; and Osceola district, White Pine County.

The available figures for placer gold production for all placer districts are given in table 1. For comparison, I have included in table 2 figures for the 24 gold districts in Nevada that have produced more than 100,000 ounces in lode gold (from Koschmann and Bergendahl, 1968).

Most of the gold recovered before 1900, an estimated 905,850 ounces, was recovered by many individuals using drywashers or small sluices to work gravels brought to the surface from shafts or pits. In the major districts (Silver City, Spring Valley, Sierra, and Osceola) worked intensely between 1849 and 1890, the miners dug numerous shafts, tunnels, and adits in the gravels. At Osceola, large banks of gravel were hydraulicked, leaving sheer cliffs of unworked gravels exposed today.

TABLE 2.—*Major gold districts in Nevada*

[From Koschmann and Bergendahl (1968)]

County and district	Lode production (ounces)	Placer production (ounces)
Clark County:		
Eldorado	101, 729	[1] 168
Searchlight	246, 997	26
Elko County:		
Gold Circle	109, 765	[1] 45
Jarbidge	217, 800	([2])
Tuscarora	100, 000	([2])
Esmeralda County:		
Goldfield	4, 194, 800	([2])
Silver Peak	568, 000	
Eureka County:		
Eureka	1, 230, 000	([2])
Lynn	[3] 390, 000	[1] 9, 000
Humboldt County:		
National	177, 000	([2])
Potosi	485, 700	([2])
Lander County:		
Bullion	[4] 146, 154	[1] 10, 373
Lincoln County:		
Delamar	217, 240	
Pioche	104, 583	
Lyon County:		
Silver City	[5] 143, 500	[5] 46, 500
Pine Grove (Wilson)	408, 000	([2])

See footnotes at end of table.

TABLE 2.—*Major gold districts in Nevada*—Continued

County and district	Lode production (ounces)	Placer production (ounces)
Mineral County:		
Aurora_____	93, 600	([2])
Nye County:		
Bullfrog_____	120, 401	([2])
Manhattan_____	280, 022	[1] 206, 340
Round Mountain_____	329, 000	[1] 208, 200
Tonopah_____	1, 880, 000	([2])
Pershing County:		
Seven Troughs_____	160, 182	([2])
Storey County:		
Comstock_____	8, 560, 000	([2])
White Pine County:		
Ely (Robinson)_____	1, 959, 659	([2])

[1] See table 1 for different estimate of placer gold production.

[2] See table 1 for placer gold production; Koschmann and Bergendahl do not list placer production separately.

[3] Carlin Mine.

[4] Most from Gold Acres mine.

[5] Production data from Bergendahl (1964); see table 1 for different estimate of placer gold production.

After 1900, drywashers, small sluices, and small concentrating machines continued to be used in placers throughout the State, but, except for the very productive first few years of drywashing at Manhattan, Round Mountain, and Battle Mountain (1906–15), the greatest part of the placer gold was recovered by large dredging operations. Figure 1 is a graphic representation of the total amount of placer gold recovered yearly in Nevada (1900–68) and the contributors to the major production peaks.

Dredge mining in Nevada started in 1911, when the Federal Mining Co. used a small wooden dredge to work gravels in Spring Valley Canyon (Pershing County). The operation was only moderately successful, but it encouraged other companies to consider desert dredge mining. During the periods 1920–23, 1940–42, and 1946–47, dredges worked in the relatively well-watered Carson River at Gold Canyon (Dayton, Lyon County). Small dredges worked gravels in a number of districts throughout the State (such as the Bullion district, Lander County; the Willow Creek district, Pershing County; and the Olinghouse district, Washoe County), but in many of these operations, the water was not sufficient for the use of floating dredges, and other conveyances were used to transport the gravels to the dredge, which acted as a central washing plant. The era of major large-scale desert dredge operations began in 1939, when a floating bucketline dredge was

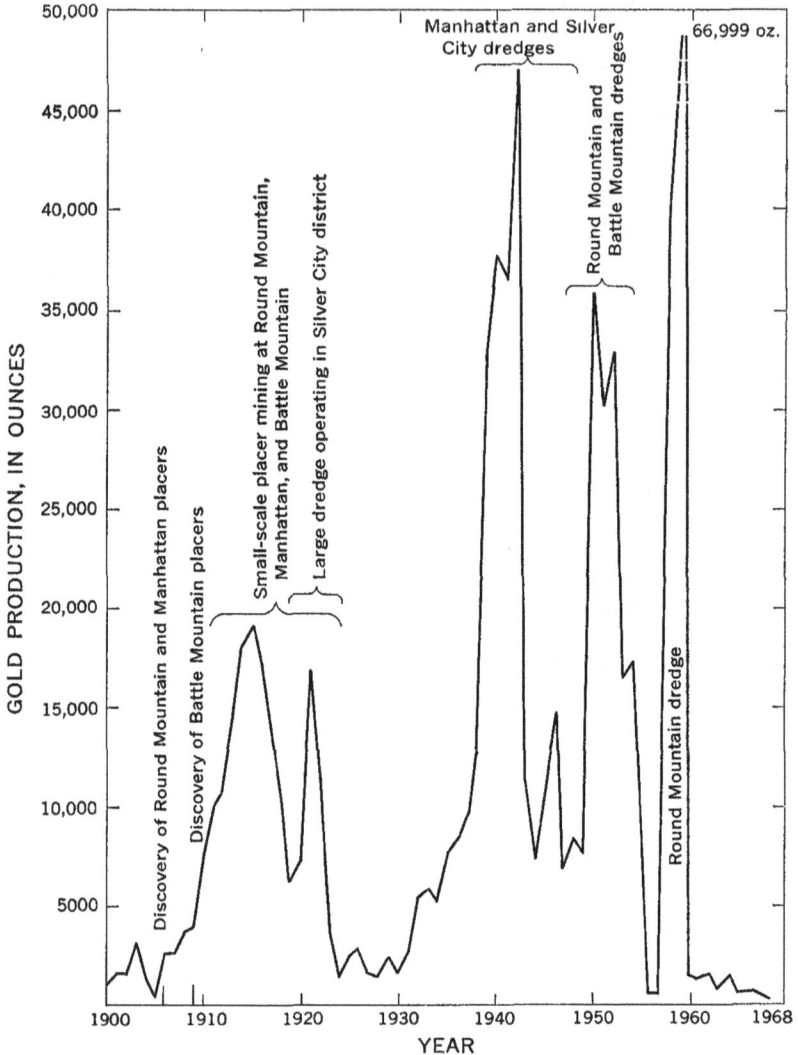

FIGURE 1.—Graph showing Nevada placer gold production, in ounces.

brought to Manhattan Gulch (Nye County). When operations ceased in 1946, this same dredge was transported to Battle Mountain (Lander County) to work the placers in the Copper Canyon fan from 1947 to 1955. In the 1950's, a nonfloating dredge was used at Round Mountain (Nye County) to recover large amounts of placer gold from a deep pit. Since cessation of dredge operations at Round Mountain in 1959, placer gold production in Nevada has returned to small-scale sporadic or part-time operations by individuals.

SUMMARY

Placer gold has been found in 115 mining districts in Nevada. Many of these districts have produced, or are said to have produced, only a few ounces of placer gold. Thirteen districts have produced more than 10,000 ounces. Although placer gold has been recovered from each of the 17 counties in Nevada, most of the placers are in the western part of the State (see pl. 1) in the area termed the "Western Metallogenic Province," which is characterized by the dominance of precious metal ores (Roberts, 1946a). A few placers (all of minor importance except the Osceola district) are found in the eastern part of the State in the area termed the "Eastern Metallogenic Province," which is characterized by the dominance of base-metal ores.

Most of the placer gold found in Nevada has been derived from veins and replacement deposits that have been successfully worked for the gold and silver content of the ores. In the few districts where source of the gold is unknown, it is presumed to be small scattered veins in the adjacent bedrock. In most of the very productive lode mining districts, only small amounts of placer gold have been recovered, whereas in the very productive placer districts, lode-gold production is close to, and sometimes less than, placer gold production. An exception is the Silver City district (Lyon County), which has yielded a high production of placer gold derived from ores of the Comstock lode (Lyon and Storey Counties), the largest silver-producing district in the State.

AGE OF LODE MINERALIZATION

The lode deposits that are the source of the placer gold range in age from ores of probable Precambrian age to ores dated as late Tertiary age.

Most of the placer deposits in Nevada were derived from ore deposits formed during the Tertiary. Unfortunately the only information available about the age of the ore deposits of many districts is that they are genetically related to intrusive and extrusive rocks of known Tertiary age. Important placer districts where the age of mineralization has been established by radiometric dating of ores or associated intrusive rocks are the Tuscarora district, Elko County, and Battle Mountain and Tenabo districts, Lander County (intrusion and subsequent mineralization occurred about 38 m.y. ago), and the Comstock-Silver City district, Storey and Lyon Counties (intrusion and subsequent mineralization occurred about 12–14 m.y. ago). The ores in the Round Mountain and Manhattan districts, Nye County, and Rawhide district, Mineral County, also formed during the Tertiary, but the exact age of mineralization is unknown.

Pre-Tertiary mineralization has formed ores that are the source of placer deposits in some districts in Nevada. Most of these districts contain relatively minor deposits of placer gold, but a few—Spring Valley, Sierra,

Island Mountain, Lida, and Mount Siegel—have produced more than 1,000 ounces.

AGE AND DISTRIBUTION OF PLACERS

Most of the placers in Nevada are in gravel accumulations in gullies, drywashes, and hillsides. The gold is found mixed with debris from the adjacent bedrock and lode deposits and is typically concentrated at or near bedrock or false bedrock. In areas where only minor amounts of placer gold have been found, either the amount of erosion of the mineralized bedrock has been minor or the gravel accumulations are a thin covering over bedrock. In areas where extensive placers have been found and worked, the placer gravels characteristically are thick accumulations of debris eroded from the mineralized bedrock and deposited in well-defined stream or gully channels, in alluvial fans at the edges of the range fronts, or in thick residual hillside debris. With very few exceptions, the placer gold recovered from gravels in Nevada has been transported only a very short distance from the source. The gold occurs as angular grains typically coarser nearer to the source than short distances downslope. For the three districts in the State where the placer gold is thought to have been transported for considerable distances—Little Valley, Washoe County; Genoa district, Douglas County; and Yerington district, Lyon County—there is very little accurate information about the source of the placers.

Most of the placers formed during erosional cycles of Quaternary age. The placer gravels in Manhattan Gulch (Manhattan district, Nye County) are Pleistocene in age and are buried under thick barren deposits of later Quaternary age. So far as can be determined, most of the placers formed during the later part of the Quaternary during recent erosional cycles and earlier cycles subsequent to the end of the comparatively well-watered Pleistocene. A few minor placers occur in gravels considered to be Tertiary in age (Little Valley, Genoa, Yerington, and Mount Siegel).

BIBLIOGRAPHY

LITERATURE REFERENCES

Albers, J. A., and Stewart, J. H., 1920, Geology and mineral deposits of Esmeralda County, Nevada: Nevada Bur. Mines Bull. (in press). Describes geology of mining districts in county.

Archbold, N. L., and Paul, R. R., 1970, Geology and mineral deposits of the Pamlico mining district, Mineral County, Nevada: Nevada Bur. Mines Bull. 74, 12 p.

Beal, L. H., 1965, Geology and mineral deposits of the Bunkerville mining district, Clark County, Nevada: Nevada Bur. Mines Bull. 63, 96 p.

Bergendahl, M. H., 1964, Gold, *in* Mineral and water resources of Nevada: Nevada Bur. Mines Bull. 65, p. 87–100.

Blake, J. W., 1964, Geology of the Bald Mountain intrusive, Ruby Mountain, Nevada: Brigham Young Univ. Geology Studies, v. 11, p. 3–35.

Bonham, H. F., 1969, Geology and mineral deposits of Washoe and Storey Counties, Nevada: Nevada Bur. Mines Bull. 70, 140 p.

Browne, J. R., 1868, Report on the mineral resources of the States and Territories west of the Rocky Mountains (for the year 1867) : Washington [U.S. Treasury Dept.], 674 p. [Nevada, p. 299–442].

Buckley, E. R., 1911, Geology of the Jarbidge mining district, Nevada: Mining and Eng. World, v. 35, p. 1209–1212.

Burchard, H. C., 1883, Report of the Director of the Mint upon the statistics of the production of the precious metals in the United States (for the year 1882): Washington, U.S. Bur. Mint, 873 p. [Nevada, p. 136–179].

———— 1884, Report of the Director of the Mint upon the production of the precious metals in the United States during the calendar year 1883: Washington, U.S. Bur. Mint, 858 p. [Nevada p. 500–561].

———— 1885, Report of the Director of the Mint upon the production of the precious metals in the United States during the calendar year 1884: Washington, U.S. Bur. Mint, 644 p. [Nevada, p. 339–372].

Cameron, E. N., 1939, Geology and mineralization of the north-eastern Humboldt Range, Nevada: Geol. Soc. America Bull., v. 50, p. 563–634.

Clark, A. N., 1946, Nevada's Manhattan gold dredge: Mining Jour. [Phoenix, Ariz.], v. 29, no. 23, p. 2–4.

———— 1947, Nevada Desert Placer mining: Mining World, v. 9, no. 8, p. 27–30.

Coash, J. R., 1967, Geology of the Mount Velma quadrangle, Elko County, Nevada: Nevada Bur. Mines Bull. 68, 20 p.

Cornwall, H. R., 1972, Geology and mineral resources of southern Nye County, Nevada: Nevada Bur. Mines Bull. (in press).

Cornwall, H. R., and Kleinhampl, F. J., 1964, Geology of Bullfrog quadrangle and ore deposits related to Bullfrog Hills, Caldera, Nye County, Nevada, and Inyo County, California: U.S. Geol. Survey Prof. Paper 454-J, p. 1–25.

Decker, R. W., 1962, Geology of the Bull Run quadrangle, Elko County, Nevada: Nevada Bur. Mines Bull. 60, 65 p.

De Quille, Dan, 1891, The discovery of the Comstock lode: Eng. and Mining Jour., v. 52, p. 637–638.

Dreyer, R. M., 1940, Goldbanks mining district, Pershing County, Nevada: Nevada Bur. Mines Bull. 33, 36 p.

Emmons, W. H., 1910, A reconnaissance of some mining camps in Elko, Lander, and Eureka Counties, Nevada: U.S. Geol. Survey Bull. 408, 130 p.

Engineering and Mining Journal, 1887, General mining news—Nevada [Osceola district]: Eng. and Mining Jour., v. 44, p. 420.

———— 1891, General mining news—Nevada [Osceola district] : Eng. and Mining Jour., v. 52, p. 133.

———— 1892a, General mining news—Nevada [Osceola district]: Eng. and Mining Jour., v. 53, p. 117.

———— 1892b, General mining news—Nevada [Osceola district]: Eng. and Mining Jour., v. 54, p. 304.

———— 1893, General mining news—Nevada [Bruneau River]: Eng. and Mining Jour., v. 56, p. 505.

———— 1896a, General mining news—Nevada [Island Mountain district]: Eng. and Mining Jour., v. 62, p. 517.

———— 1896b. General mining news—Nevada [Mount Siegel district]: Eng. and Mining Jour., v. 61, p. 46.

———— 1897a, General mining news—Nevada [Alder district]: Eng. and Mining Jour., v. 64, p. 47.

———— 1897b, General mining news—Nevada [Island Mountain district]: Eng. and Mining Jour., v. 64, p. 227.

———— 1897c, General mining news—Nevada [Island Mountain district]: Eng. and Mining Jour., v. 64, p. 737.

———— 1897d, General mining news—Nevada [Olinghouse district]: Eng. and Mining Jour., v. 64, p. 288.

———— 1898, General mining news—Nevada [Island Mountain district]: Eng. and Mining Jour., v. 65, p. 292.

———— 1916, Round Mountain placers: Eng. and Mining Jour., v. 101, p. 856.

———— 1921, News by mining districts—Nevada [Johnnie district]: Eng. and Mining Jour., v. 112, p. 633.

———— 1931, Brief notes—Nevada-Montana starts plant [Placerites district]: Eng. and Mining Jour., v. 132, no. 3, p. 134.

———— 1933a, Trends & Developments in the industry—Nevada [Buckskin district]: Eng. and Mining Jour., v. 134, p. 387.

———— 1933b, Trends & Developments in the industry—Nevada [Rabbit Hole district]: Eng. and Mining Jour., v. 134, no. 3, p. 130.

———— 1942, Trends & Developments in the industry—Nevada [Rabbit Hole district]: Eng. and Mining Jour., v. 143, no. 4, p. 93; no. 5, p. 78.

———— 1950, In the United States—Nevada [Hornsilver district]: Eng. and Mining Jour., v. 151, no. 8, p. 144.

———— 1958a, In the United States—Nevada [Ione district]: Eng. and Mining Jour., v. 159, no. 11, p. 168.

———— 1958b, In the United States—Nevada [Round Mountain district]: Eng. and Mining Jour., v. 159, no. 3, p. 182.

Ferguson, H. G., 1917, Placer deposits of the Manhattan district, Nevada: U.S. Geol. Survey Bull. 640, p. 163–193.

———— 1922, The Round Mountain district, Nevada: U.S. Geol. Survey Bull. 725, p. 383–406.

———— 1924, Geology and ore deposits of Manhattan district, Nevada: U.S. Geol. Survey Bull. 723, p. 117–133.

———— 1927, The Gilbert district, Nevada: U.S. Geol. Survey Bull. 795–F, p. 125–145.

———— 1944, Mining districts of Nevada: Nevada Bur. Mines Bull. 40, p. 77–108.

Ferguson, H. G., and Cathcart, S. H., 1954, Geology of the Round Mountain quadrangle, Nevada: U.S. Geol. Survey Geol. Quad. Map GQ–40, scale 1:125,000, with text.

Ferguson, H. G., and Muller, S. W., and Cathcart, S. H., 1953, Geology of the Coaldale quadrangle, Nevada: U.S. Geol. Survey Geol. Quad. May GQ–23, scale 1:125,000.

Ferguson, H. G., Muller, S.W., and Roberts, R. J., 1951, Geology of the Winnemucca quadrangle, Nevada: U.S. Geol. Survey Geol. Quad. Map GQ–11, scale 1:125,000.

Gardner, E. D., and Allsman, P. T., 1938, Power shovel and dragline placer mining: U.S. Bur. Mines Inf. Circ. 7013, 68 p.

Gianella, V. P., 1936, Geology of the Silver City district and the southern portion of the Comstock lode, Nevada: Nevada Bur. Mines Bull. 30, 105 p.

Granger, A. E., Bell, M. M., Simmons, G. C., and Lee, Florence, 1957, Geology and mineral resources of Elko County, Nevada: Nevada Bur. Mines Bull. 54, 190 p.

Summarizes earlier reports about the geology of mining districts in Elko County; includes reference list to major publications for each district; describes some of the placers in the county.

Gray, R. F., 1951, Geology of a portion of the Pine Nut Mountains, Nevada: California Univ. (Berkeley), M.A. thesis, 75 p.

Hill, J. M., 1911, Notes on the economic geology of the Ramsey, Talapoosa and Whitehorse mining districts in Lyon and Washoe Counties, Nevada: U.S. Geol. Survey Bull. 470, p. 99–108.

———— 1915, Some mining districts in northeastern California and northwestern Nevada: U.S. Geol. Survey Bull. 594, 200 p.

———— 1916, Notes on some mining districts in eastern Nevada: U.S. Geol. Survey Bull. 648, 214 p.

Hose, R. K., Blake, M. C., Jr., and Smith, R. M., 1972, Geology and mineral resources of White Pine County, Nevada: Nevada Bur. Mines Bull. (in press).

Hotz, P. E., and Willden, Ronald, 1964, Geology and mineral resources of the Osgood Mountains quadrangle, Humboldt County, Nevada: U.S. Geol. Survey Prof. Paper 431, 128 p.

Huttl, J. B., 1934, Portable Washing plant: Eng. and Mining Jour., v. 135, no. 4, p. 173.

———— 1950a, How Natomas keeps a large dredge operating in the desert: Eng. and Mining Jour., v. 151, no. 10, p. 96–99.

———— 1950b, New 17,000 ton Dry-land "dredge" used draglines, shovel, belts, washing plant: Eng. and Mining Jour., v. 151, no. 6, p. 68–70.

Jones, C. C., 1909, Notes on Manhattan placers, Nye County, Nevada: Eng. and Mining Jour., v. 88, p. 101–104.

Jones J. C., Smith, A. M., and Stoddard, Carl, 1931, The preliminary survey of the Scossa mining district, Pershing County, Nevada: Nevada Bur. Mines Bull. 11, 14 p.

Knopf, Adolph, 1924, Geology and ore deposits of the Rochester district, Nevada: U.S. Geol. Survey Bull. 762, 78 p.

Koschmann, A. H., and Bergendahl, M. H., 1968, Principal gold-producing districts of the United States: U.S. Geol. Survey Prof. Paper 610, 283 p.

Kral, V. E., 1951, Mineral resources of Nye County, Nevada: Nevada Bur. Mines Bull. 50, 220 p.

Labbe, Charles, 1921, The placers of the Johnnie district, Nevada: Eng. and Mining Jour., v. 112, p. 895–896.

Lincoln, F. C., 1923, Mining districts and mineral resources of Nevada: Reno, Nevada Newsletter Publishing Co., 295 p.

Lindgren, Waldemar, 1915, Geology and mineral deposits of the National mining district, Nevada: U.S. Geol. Survey Bull. 601, 58 p.

Locke, E. G., 1913, The rewakening of an old placer camp: Mining and Sci. Press, v. 106, p. 373.

Longwell, C. R., Pampeyan, E. H., Bowyer, Ben, and Roberts, R. J., 1965, Geology and mineral deposits of Clark County, Nevada: Nevada Bur. Mines Bull. 62, 218 p.

Lord, Eliot, 1883, Comstock mining and miners: U.S. Geol. Survey Mon. 4, 451 p.

Luther, L. A., 1950, Panning gold with 1500 horsepower: Compressed Air Mag., v. 55, p. 232–236.

Martin, A. H.., 1910, The Bannock mining district, Nevada: Mining World, v. 32, p. 835.

———— 1912, Present status of the Manhattan district, Nevada: Mining Sci., v. 65, p. 248–249.

Martin, Gail, 1931, Tuscarora responds to modern mine methods: Mining Jour. [Phoenix, Ariz.], v. 14, no. 23, p. 3–4.

Mining and Engineering World, 1913a, Ingenious placer operations near Manhattan, Nevada: Mining and Eng. World, v. 39, p. 200.

———— 1913b, Late news from the World's mining camps—Nevada [Jumbo district]: Mining and Eng. World, v. 39, p. 807.

Mining and Scientific Press, 1908a, General mining news—Nevada [Centennial district]: Mining and Sci. Press, v. 96, p. 214.

———— 1908b, Rawhide, Nevada: Mining and Sci. Press, v. 96, p. 281.

———— 1908c, Rawhide, Nevada: Mining and Sci. Press, v. 96, p. 382–383.

———— 1908d, General mining news—Nevada [Round Mountain]: Mining and Sci. Press, v. 96, p. 792.

———— 1909, General mining news—Nevada [Ione district]: Mining and Sci. Press, v. 98, p. 205.

Mining Journal, 1928, Western Mining news gathered in Los Angeles—[Cloverdale Canyon placers]: Mining Jour. [Phoenix, Ariz.], v. 11, no. 19, p. 3.

———— 1931, Pacific Coast mining activities—Nevada [Rochester district]: Mining Jour. [Phoenix, Ariz.], v. 15, no. 4, p. 28.

———— 1938a, Concentrates from the Western States—Nevada [Jessup district]: Mining Jour. [Phoenix, Ariz.], v. 22, no. 14, p. 30.

———— 1938b, Concentrates from the southwest—Nevada [Antelope district]: Mining Jour. [Phoenix, Ariz.], v. 22, no. 10, p. 20.

———— 1939a, Mill Heads from the Western States—Nevada [Willow Creek]: Mining Jour [Phoenix, Ariz.], v. 22, no. 22, p. 25.

———— 1939b, Rabbit Hole area governed by old time mining rules: Mining Jour. [Phoenix, Ariz.], v. 23, no. 5, p. 34.

———— 1940a, Nuggets from the Western States—Nevada [Rabbit Hole district]: Mining Jour. [Phoenix, Ariz.], v. 24, no. 3, p. 25.

———— 1940b, Nuggets from the Western States—Nevada [Sierra district]: Mining Jour. [Phoenix, Ariz.], v. 24, no. 3, p. 24.

———— 1945, Nuggets from the Western States—Nevada [Yerington district]: Mining Jour. [Phoenix, Ariz.], v. 29, no. 4, p. 23.

Mining Review, 1910, To work Osceola placers: Mining Rev. [Salt Lake City, Utah], v. 12, no. 5, p. 38.

———— 1923, Another placer gravel discovery [Carson River]: Mining Rev. [Salt Lake City, Utah], v. 25, no. 13, p. 17.

———— 1933, Lots of things happening in the Bald Mountain district, Nevada: Mining Rev. [Salt Lake City, Utah], v. 35, no. 44, p. 9.

Mining World, 1907a, Late news from the World's mining camps—Nevada [Centennial district]: Mining World, v. 27, p. 825.

———— 1907b, Late news from the World's mining camps—Nevada [Tuscarora]: Mining World, v. 27, p. 945.

———— 1908, Late news from the World's mining camps—Nevada [Round Mountain]: Mining World, v. 28, p. 270.

———— 1909, Late news from the World's mining camps—Nevada [Las Vegas Wash]: Mining World, v. 30, p. 855.

———— 1910, Late news from the World's mining camps—Nevada [Rabbit Hole district]: Mining World, v. 33, p. 292.

———— 1911, Late news from the World's mining camps—Nevada [Gold Run district]: Mining World, v. 24, p. 460.

—— 1940, Natomas Company secures the Copper Canyon placers: Mining World, v. 2, no. 3, p. 20.

—— 1941a, Dayton, Nevada—dragline dredging hits all-time high: Mining World, v. 3, no. 2, p. 7–16.

—— 1941b, Manhattan—where stripping constitutes a major element in gold dredging operations: Mining World, v. 3, no. 3, p. 17–23.

—— 1947, Precipates—Southwest—Nevada [Millett district]: Mining World, v. 9, no. 8, p. 70.

—— 1950, Round Mountain Gold: Mining World, v. 12, no. 6, p. 26–31.

—— 1951, The Round Mountain Mill: Mining World, v. 13, no. 9, p. 20–23.

—— 1959, Mining World Newsletter [Round Mountain district]: Mining World, v. 21, no. 5, p. 7

Moore, J. G., 1969, Geology and mineral deposits of Lyon, Douglas, and Ormsby Counties, Nevada: Nevada Bur. Mines Bull. 75, 45 p.

Murbarger, Nell, 1957, Only the sidewalk remains at Gold Creek: Desert Mag., v. 20, no. 2, p. 17–22.

—— 1958, Chinese ghost town in the Humboldt Range: Desert Mag., v. 21, no. 11, p. 4–7.

Nevada Bureau of Mines, 1962, Geological, geophysical, and hydrological investigations of the Sand Springs Range, Fairview Valley, and Fourmile Flat, Churchill County, Nevada, for Shoal Event Project Shade, Vela Uniform Program, Atomic Energy Commission: Reno, Nev., Univ. Nevada, 127 p.

Nevada Mining Press, 1929, Impound water to sluice auriferous gravel deposit on desert, Pershing County: Nevada Mining Press [newspaper], June 28, 1929, v. 11, no. 474, p. 4.

—— 1930a, Install dry washing plant at Rawhide to recover placer gold: Nevada Mining Press [newspaper], Nov. 14, 1930, v. 13, no. 546, p. 1.

—— 1930b, Idaho Gold Dredging Co. proposes to dredge Rawhide placers: Nevada Mining Press [newspaper], Dec. 12, 1930, v. 13, no. 550, p. 1.

—— 1931a, Work of sampling placer deposits at Rawhide stops: Nevada Mining Press [newspaper], May 8, 1931, v. 13, no. 571, p. 4.

—— 1931b, Mining men rush to new placer discovery, head of Smith Valley: Nevada Mining Press, [newspaper], May 15, 1931, v. 13, no. 572, p. 1.

—— 1931c, Smith Valley placer remmant of old river say State Engineers: Nevada Mining Press [newspaper], May 22, 1931, v. 13, no. 573, p. 1, 4.

Nolan, T. B., 1936a, Nonferrous metal deposits, in Hewett, D. F., and others, Mineral resources of the region around Boulder Dam: U.S. Geol. Survey Bull. 871, p. 5–77.

—— 1936b, The Tuscarora mining district, Elko County, Nevada: Nevada Bur. Mines, v. 25, 36 p.

—— 1962, The Eureka mining district, Nevada: U.S. Geol. Survey Prof. Paper 406, 78 p.

Overton, T. D., 1947, Mineral resources of Douglas, Ormsby and Washoe Counties: Nevada Bur. Mines Bull. 46, 88 p.

Packard, G. A., 1907, Round Mountain Camp, Nevada: Eng. and Mining Jour., v. 83, p. 150–151.

—— 1908, Round Mountain, Nevada: Mining and Sci. Press, v. 96, p. 807–809.

Page, B. M., 1959, Geology of the Candelaria mining district, Mineral County, Nevada: Nevada Bur. Mines Bull. 56, 67 p.

Paher, S. W., 1970, Nevada ghost towns and mining camps: Berkeley, Calif., Howell-North Brooks, 492 p.

Location and history of nearly 600 mining towns in Nevada; briefly mentions placer-mining history for some areas. Includes photographs of various placer-mining operations throughout the state.

Penrose, R. J., 1937, Singatse channel in Nevada: Mining Jour. [Phoenix, Ariz.], v. 21, no. 6, p. 3–4.

Ransome, F. L., 1909a, The Hornsilver district, Nevada: U.S. Geol. Survey Bull. 380, p. 41–43.

―――― 1909l, Notes on some mining districts in Humboldt County, Nevada: U.S. Geol. Survey Bull. 414, 75 p.

―――― 1909c, Round Mountain, Nevada: U.S. Geol. Survey Bull. 380, p. 44–47.

Raymond, R. W., 1870, Statistics of mines and mining in the States and Territories west of the Rocky Mountains for the year 1869: Washington, U.S. Treasury Dept., 805 p. [Nevada, p. 89–201].

―――― 1872, Statistics of mines and mining in the States and Territories west of the Rocky Mountains for the year 1870: Washington, U.S. Treasury Dept., 566 p. [Nevada, p. 93–175].

―――― 1873, Statistics of mines and mining in the States and Territories west of the Rocky Mountains for the year 1871: Washington, U.S. Treasury Dept., 566 p. [Nevada, p. 141–249].

―――― 1877, Statistics of mines and mining in the States and Territories west of the Rocky Mountains for the year 1875: Washington, U.S. Treasury Dept., 519 p. [Nevada, p. 132–202].

Reid, J. A., 1908, A tertiary river channel near Carson City, Nevada: Mining and Sci. Press, v. 96, p. 522–525.

Richardson, J. W., 1936, Placer gravels in the Colorado River Basin: Mining Jour. [Phoenix, Ariz.], v. 19, no. 22, p. 4.

Roberts, R. J., 1964a, Economic geology, in Mineral and water resources of Nevada: Nevada Bur. Mines, Bull. 65, p. 39–48.

Summarizes characteristics of ore deposits; age and alinement of mineral districts.

―――― 1964b, Stratigraphy and structure of the Antler Peak quadrangle, Humboldt and Lander Counties, Nevada: U.S. Geol. Survey Prof. Paper 459–A, 93 p.

Roberts, R. J., and Arnold, D. C., 1965, Ore deposits of the Antler Peak quadrangle, Humboldt and Lander Counties, Nevada: U.S. Geol. Survey Prof. Paper 459–B, 93 p.

Roberts, R. J., Montgomery, K. M., and Lehner, R. E., 1967, Geology and mineral resources of Eureka County, Nevada: Nevada Bur. Mines Bull. 64, 152 p.

Roberts, R. J., Radtke, A. S., and Coats, R. R., 1971, Gold-bearing deposits in north-central Nevada and southwestern Idaho: Econ. Geology, v. 66, no. 1, p. 14–33.

Summarizes major characteristics, ages, and mineralogy of many important gold lode deposits.

Ross, C. P., 1953, The geology and ore deposits of the Reese River district, Lander County, Nevada: U.S. Geol. Survey Bull. 997, 132 p.

Ross, D. C., 1961, Geology and mineral deposits of Mineral County, Nevada: Nevada Bur. Mines Bull. 58, 98 p.

Rott, E. H., Jr., 1931, Ore deposits of the Gold Circle mining district, Elko County, Nevada: Nevada Bur. Mines Bull. 12, 29 p.

Schrader, F. C., 1912, A reconnaissance of the Jarbidge, Contact, and Elk Mountain mining districts, Elko County, Nevada: U.S. Geol. Survey Bull. 497, 162 p.

——— 1915, The Rochester mining district, Nevada: U.S. Geol. Survey Bull. 580, p. 325–372.

——— 1923, The Jarbidge mining district, Nevada, with a note on the Charleston district: U.S. Geol. Survey Bull. 741, 86 p.

——— 1934, The McCoy mining district and gold veins in Horse Canyon Lander County, Nevada: U.S. Geol. Survey Circ. 10, 13 p.

——— 1947, Mining districts in the Carson Sink region, Nevada: U.S. Geol. Survey open-file report, 523 p., 100 maps.

Silberman, M. L., and McKee, E. H., 1971, Periods of plutonism in north-central Nevada, in Roberts, R. J., Radtke, A. S., and Coats, R. R., Gold-bearing deposits in north-central Nevada and southwestern Idaho: Econ. Geology, v. 66, no. 1, p. 17–19.

Silberman, M. L., Wrucke, C. T., and Armbrustmacher, T. J., 1969, Age of mineralization and intrusive relations at Tenabo, northern Shoshone Range, Lander County, Nevada [abs.]: Geol. Soc. America, Cordilleran Sec.—Paleont. Soc., Pacific Coast Sec., 65th Ann. Mtg., Eugene, Oreg., 1969, Program, pt. 3, p. 62.

Smith, A. M., 1932, The Mountain City mining district of Nevada: Mining Jour. [Phoenix, Ariz.], v. 16, no. 13, p. 5–6.

Smith, A. M., and Stoddard, Carl, 1932, Mines in northeastern Nevada: Mining Jour. [Phoenix, Ariz.], v. 16, no. 4, p. 3–4.

Smith, A. M., and Vanderburg, W. O., 1932, Placer mining in Nevada: Nevada Bur. Mines Bull. 18, 104 p.
Summary of present (1932) status of placer gold operations. Methods of mining are treated extensively.

Southern Pacific Company, 1964, Minerals for industry—Northern Nevada and northwestern Utah, summary of Geol. Survey of 1955–1961, volume 1: San Francisco, Calif., Southern Pacific Company, 188 p.

Stewart, J. H., and McKee, E. H., 1968, Favorable areas for prospecting adjacent to the Roberts Mountains thrust in southern Lander County, Nevada: U.S. Geol. Survey Circ. 563, 13 p.

Stoddard, Carl, and Carpenter, J. A., 1950, Mineral resources of Storey and Lyon Counties, Nevada: Nevada Bur. Mines Bull. 49, 111 p.

Stoneham, W. J., 1911, Manhattan placers, Nevada: Mining and Eng. World, v. 35, p. 242.

Stuart, E. E., 1909, Nevada's mineral resources: Carson City, Nev., State Printing Office, 158 p.

Theodore, T. G., and Roberts, R. J., 1971, Geochemistry and geology of deep drill holes at Iron Canyon, Lander County, Nevada: U.S. Geol. Survey Bull. 1318, 32 p.

Thompson, G. A., 1956, Geology of the Virginia City quadrangle, Nevada: U.S. Geol. Survey Bull. 1042–C, p. 45–77.
Notes extent of placers; describes lode mines.

Thompson, G. A., and White, D. E., 1964, Regional geology of the Steamboat Springs area, Washoe County, Nevada: U.S. Geol. Survey Prof. Paper 458–A, 52 p.

Toll, R. H., 1911, Present aspect of the Manhattan district, Nevada: Mining and Eng. World, v. 35, p. 639–640.

Tonopah Times-Bonanza [Newspaper], 1970, Friday August 21, Round Mountain joint venture set [by Nettie Darrough]: Tonopah Times-Bonanza.

——— 1972, Friday, Feb. 11, Cooper Range expands test programs: Tonopah Times-Bonanza.

Tschanz, C. M., and Pampeyan, E. H., 1970, Geology and mineral deposits of Lincoln County, Nevada: Nevada Bur. Mines Bull. 73, 187 p.

U.S. Bureau of Mines, 1967, Production potential of known gold deposits in the United States: U.S. Bur. Mines Inf. Circ. 8331, 24 p.

———— 1925–34, Mineral resources of the United States [annual volumes, 1924–31]: Washington, U.S. Govt. Printing Office.

———— 1933–68, Minerals Yearbook [annual volumes, 1932–68]: Washington, U.S. Govt. Printing Office.

U.S. Geological Survey, 1896–1900, Annual reports [17th through 21st, 1895–1900]: Washington, U.S. Govt. Printing Office.

———— 1883–1924, Mineral resources of the United States [annual volumes, 1882–1923]: Washington, U.S. Govt. Printing Office.

———— 1969, U.S. Geological Survey Heavy Metals Program progress report 1968 —field studies: U.S. Geol. Survey Circ. 621, 35 p. [Shawe, D. R., and Poole, F. G., p. 23, Manhattan mineral belt.]

Vanderburg, W. O., 1936a, Placer mining in Nevada: Nevada Bur. Mines Bull. 30, 178 p.
    Revised discussion of placer operations. Includes three localities previously unreported.

———— 1936b, Reconnaissance of mining districts in Pershing County, Nevada: U.S. Bur. Mines Inf. Circ. 6902, 57 p.

———— 1937a, Reconnaissance of mining districts in Clark County, Nevada: U.S. Bur. Mines Inf. Circ. 6964, 81 p.

———— 1937b, Reconnaissance of mining districts in Mineral County, Nevada: U.S. Bur. Mines Inf. Circ. 6941, 79 p.

———— 1938a, Reconnaissance of mining districts in Eureka County, Nevada: U.S. Bur. Mines Inf. Circ. 7022, 66 p.

———— 1938b, Reconnaissance of mining districts in Humboldt County, Nevada: U.S. Bur. Mines Inf. Circ. 6995, 54 p.

———— 1939, Reconnaissance of mining districts in Lander County, Nevada: U.S. Bur. Mines Inf. Circ. 7043, 83 p.

———— 1940, Reconnaissance of mining districts in Churchill County, Nevada: U.S. Bur. Mines Inf. Circ. 7093, 57 p.

Walker, H. G., 1911, First gold dredge in Nevada: Eng. and Mining Jour., v. 91, p. 1210–1211.

Wallace, R. E., and Tatlock, D. B., 1962, Suggestions for prospecting in the Humboldt Range and adjacent areas, Nevada: U.S. Geol. Survey Prof. Paper 450–B, p. B3–B5.

Weeks, F. B., 1908, Geology and mineral resources of the Osceola mining district, White Pine County, Nevada: U.S. Geol. Survey Bull. 340, p. 117–133.

White, A. F., 1871, Third biennial report of the State Mineralogist for the years 1869–70: Carson City, Nev., 115 p.

Whitehill, H. R. [1873], Biennial report of the State Mineralogist of the State of Nevada for the years 1871 and 1872: Carson City, Nev., Concurrent Resolution (Printers), 191 p.

———— [1875], Biennial report of the State mineralogist of the State of Nevada for the years 1873 and 1874: Carson City, Nev., State Printer, John J. Hill, 191 p.

———— 1877, Biennial report of the State Mineralogist of the State of Nevada for the years 1875 and 1876: Carson City, Nev., State Printer, John J. Hill. 226 p.

———— 1879, Biennial report of the State Mineralogist of the State of Nevada for the years 1877 and 1878: San Francisco, Calif., A. L. Bancroft & Co., 212 p.

Willden, Ronald, 1964, Geology and mineral deposits of Humboldt County, Nevada: Nevada Bur. Mines Bull. 59, 154 p.

Willden, Ronald, and Hotz, P. E., 1955, A gold-scheelite-cinnabar placer in Humboldt County, Nevada: Econ. Geology, v. 50, no. 7, p. 661–668.

Willden, C. R., and Speed, R. C., 1968, Geology and mineral deposits of Churchill County, Nevada: U.S. Geol. Survey open-file report.

Describes geology of mining districts in Churchill County; does not describe small placers in county.

Wolcott, G. E., 1909, Mining and milling at Rawhide, Nevada: Eng. and Mining Jour., v. 87, p. 345–348.

Wrucke, C. T., Armbrustmacher, T. J., and Hessin, T. D., 1968, Distribution of gold, silver, and other metals near Gold Acres and Tenabo, Lander County, Nevada: U.S. Geol. Survey Circ. 589, 19 p.

York, Bernard, 1944, Geology of Nevada ore deposits: Nevada Bur. Mines Bull. 40, p. 1–76.

Formation of placer deposits are discussed on pages 22–24. Gold placers in Nevada are found on hillside alluvium or in channels of intermittent streams. Few gold placers are found in well-defined channels of permanent streams.

Young, G. J., 1920, Gold, dredging started at Dayton, Nevada: Eng. and Mining Jour., v. 110, p. 640.

———— 1921, Dredge construction at Dayton, Nevada: Eng. and Mining Jour., v. 112, p. 91–96.

GEOLOGIC MAP REFERENCES

[References keyed by number to districts given in text]

Archbold, N. L., and Paul, R. R., 1970, Geology and mineral deposits of the Pamlico Mining district, Mineral County, Nevada: Nevada Bur. Mines, Bull. 74, 12 p. pls. 1, 2, scale 1:24,000.
No. 65.

Bonham, H. F., 1969, Geology and mineral deposits of Washoe and Storey Counties, Nevada: Nevada Bur. Mines Bull. 70, 140 p., map scale 1:250,000.
Nos. 107, 108.

Coash, J. R., 1967, Geology of the Mount Velma quadrangle, Elko County, Nevada: Nevada Bur. Mines Bull. 68, pl. 1, horizontal scale 1:62,500.
Nos. 15, 19.

Coats, R. R., 1964, Geology of the Jarbidge quadrangle, Nevada-Idaho: U.S. Geol. Survey Bull. 1141–M, pl. 1, scale 1:62,500.
No. 18.

———— 1968a, Preliminary geologic map of the southwestern part of the Mountain City quadrangle, Elko County, Nevada: U.S. Geol. Survey open-file map, scale 1:20,000.
No. 16.

———— 1968b, Preliminary geologic maps of the Owyhee quadrangle, Nevada: U.S. Geol. Survey open-file map, scale 1:31,680.
No. 21.

Cornwall, H. R., 1967, Preliminary geologic map of southern Nye County, Nevada: U.S. Geol. Survey open-file map, scale 1:250,000, 6 p.; Nevada Bur. Mines Bull. (in press).
No. 74.

Cornwall, H. R., and Kleinhampl, F. J., 1961, Geology of the Bare Mountain
    quadrangle, Nevada: U.S. Geol. Survey Geol. Quad. Map GQ–157, scale
    1:62,500.
    No. 73.

–––––– 1964, Geology of Bullfrog quadrangle and ore deposits related to Bull-
    frog Hills caldera, Nye County, Nevada, and Inyo County, California: U.S.
    Geol. Survey Prof. Paper 454–J, pl. 1, scale 1:48,000.
    No. 73.

Decker, R. W., 1962, Geology of the Bull Run quadrangle, Elko County, Nevada:
    Nevada Bur. Mines Bull. 60, pl. 1, scale 1:62,500.
    No. 14.

Ferguson, H. G., 1917, Placer deposits of the Manhattan district, Nevada: U.S.
    Geol. Survey Bull. 640, pl. 6, scale 1:48,000.
    No. 78.

Ferguson, H. G., and Cathcart, S. H., 1954, Geology of the Round Mountain
    quadrangle, Nevada: U.S. Geol. Survey Geol. Quad. Map GQ–40, scale
    1:125,000.
    Nos. 77, 79.

Ferguson, H. G., Muller, S. W., and Roberts, R. J., 1951a, Geologic map of the
    Mount Moses quadrangle, Nevada: U.S. Geol. Survey Geol. Quad. Map GQ–
    12, scale 1:125,000.
    No. 49.

–––––– 1951b, Geology of the Winnemucca quadrangle, Nevada: U.S. Geol. Survey
    Geol. Quad. Map GQ–11, scale 1:125,000.
    Nos. 97, 98.

Gilluly, James, 1967, Geologic map of the Winnemucca quadrangle, Pershing and
    Humboldt Counties, Nevada: U.S. Geol. Survey Geol. Quad. Map GQ–656,
    scale 1:62,500.
    No. 36.

Gilluly, James, and Gates, Olcutt, 1965, Tectonic and igneous geology of the
    northern Shoshone Range, Nevada: U.S. Geol. Survey Prof. Paper 465, pl. 1,
    scale 1:31,680.
    Nos. 47, 48.

Granger, A. E., Bell, M. M., Simmons, G. C., and Lee, F., 1957, Geology and
    mineral resources of Elko County, Nevada: Nevada Bur. Mines Bull. 54, pl.
    1, scale ~ 1:250,000.
    Nos. 15, 20, 21.

Hose, R. K., and Blake, M. C., Jr., 1970, Geologic map of White Pine County,
    Nevada: U.S. Geol. Survey open-file map, scale 1:250,000.
    Nos. 112, 113.

Hotz, P. E., and Willden, Ronald, 1964, Geology and mineral resources of the
    Osgood Mountains quadrangle, Humboldt County, Nevada: U.S. Geol. Survey
    Prof. Paper 431, pl. 1, scale 1:62,500; fig. 15.
    No. 35.

Kleinhampl, F. J., and Ziony, J. I., 1967, Preliminary geologic map of northern
    Nye County, Nevada: U.S. Geol. Survey open-file map, scale 1:200,000.
    No. 75.

Knopf, Adolph, 1918, Geology and ore deposits of the Yerington district, Nevada:
    U.S. Geol. Survey Prof. Paper 114, pl. 1, scale 1:24,000.
    No. 58.

Lipman, P. W., Quinlivan, W. D., Carr, W. J., and Anderson, R. E., 1966, Geologic map of the Thirsty Canyon SE quadrangle, Nye County, Nevada: U.S. Geol. Survey Geol. Quad. Map GQ–489, scale 1:24,000.
No. 73.

Longwell, C. R., 1963, Reconnaissance geology between Lake Mead and Davis Dam, Arizona-Nevada: U.S. Geol. Survey Prof. Paper 374–E, pl. 1, scale 1:125,000.
Nos. 4, 7.

Longwell, C. R., Pampeyan, E. H., Bowyer, Ben, and Roberts, R. J., 1965, Geology and mineral deposits of Clark County, Nevada: Nevada Bur. Mines Bull. 62, pl. 12, scale 1:250,000.
Nos. 5, 6.

McKee, E. H., 1968, Geology of the Magruder Mountain area, Nevada-California: U.S. Geol. Survey Bull. 1251–H, pl. 1, scale 1:62,500.
Nos. 24, 25.

Moore, J. G., 1969, Geology and mineral resources of Lyon, Douglas, and Ormsby Counties, Nevada: Nevada Bur. Mines Bull. 75, pl. 1, scale 1:250,000.
Nos. 11, 12, 57, 58, 59, 105.

Nevada Bureau of Mines, 1962, Geological, geophysical, and hydrological investigations of the Sand Springs Range, Fairview Valley, and Fourmile Flat, Churchill County, Nevada, for Shoal Event Project Shade, Vela Uniform Program, Atomic Energy Commission: Nevada Univ., Reno, Nev., pl. 4, scale 1:31,680.
No. 3.

Nolan, T. B., 1936, The Tuscarora mining district, Elko County, Nevada: Nevada Bur. Mines Bull. 25, pl. 1, scale 1¾ in.=2,000 feet.
No. 20.

Rigby, J. K., 1960, Geology of the Buck Mountain-Bald Mountain area, southern Ruby Mountain, White Pine County, Nevada, in Boettcher, J. W., and Slian, W. W., Jr., eds., Guidebook to the geology of east-central Nevada: Intermountain Assoc. Petroleum Geologists and Eastern Nevada Geol. Soc., 11th Ann. Field Conf., Salt Lake City, Utah, 1960, p. 173–180; (fig. 5) scale ~ 1:125,000.
No. 111.

Roberts, R. J., and Arnold, D. C., 1965, Ore deposits of the Antler Peak quadrangle, Humboldt and Lander Counties, Nevada: U.S. Geol. Survey Prof. Paper 459–B, pl. 1, scale 1:62,500; pl. 3, scale 1:31,680; pl. 19.
No. 46.

Roberts, R. J., Montgomery, K. M., and Lehner, R. E., 1967, Geology and mineral resources of Eureka County, Nevada: Nevada Bur. Mines Bull. 64, pls. 3 and 7, scale 1:250,000, scale 1 in.=200 ft.
No. 31.

Ross, D. C., 1961, Geology and mineral deposits of Mineral County, Nevada: Nevada Bur. Mines Bull. 58, pl. 2, scale 1:250,000.
Nos. 64, 66.

Rott, E. H., Jr., 1931, Ore deposits of the Gold Circle mining district, Elko County, Nevada: Nevada Bur. Mines Bull. 12, pl. 1, scale 1 in.=1,000 feet.
No. 17.

Silberling, N. J., 1959, Pre-Tertiary stratigraphy and upper Triassic paleontology of the Union district, Shoshone Mountains, Nevada: U.S. Geol. Survey Prof. Paper 322, pl. 10, scale 1:24,000.
No. 76.

Silberling, N. J., and Wallace, R. E., 1967, Geologic map of the Imlay quad-
rangle, Pershing County, Nevada: U.S. Geol. Survey Geol. Quad. Map GQ–
666, scale 1:62,500.
No. 93.
Smith, J. G., 1972, Geologic map of the Duffer Peak quadrangle, Humboldt
County, Nevada: U.S. Geol. Survey Misc. Geol. Inv. Map I-606 (in press.)
scale 1:48,000.
No. 38.
Tatlock, D. B., 1969, Geologic map of Pershing County, Nevada: U.S. Geol.
Survey open-file map, scale 1:200,000.
Nos. 34, 88–92.
Thompson, G. A., 1956, Geology of the Virginia City quadrangle, Nevada: U.S.
Geol. Survey Bull. 1042–C, pl. 3, scale 1:62,500.
Nos. 59, 104, 106.
Thompson, G. A., and White, D. E., 1964, Regional geology of the Steamboat
Springs area, Washoe County, Nevada: U.S. Geol. Survey Prof. Paper 458–A,
pl. 1.
Nos. 109, 110.
Vitaliano, C. J., 1963, Cenozoic geology and sections of the Ione quadrangle, Nye
County, Nevada: U.S. Geol. Survey Mineral Inv. Field Studies Map MF–255,
scale 1:62,500.
No. 76.
Wallace, R. E., Tatlock, D. B., Silberling, N. J., and Irwin, W. P., 1969, Geologic
map of the Unionville quadrangle, Pershing County, Nevada: U.S. Geol.
Survey Geol. Quad. Map GQ–820, scale 1:62,500.
Nos. 94–96.
Willden, Ronald, 1964, Geology and mineral deposits of Humboldt County, Nevada:
Nevada Bur. Mines Bull. 59, pl. 1, scale 1:250,000.
Nos. 34, 37, 38.
Willden, C. R., and Speed, R. C., 1968, Geology and mineral deposits of Church-
ill County, Nevada: U.S. Geol. Survey open-file report, pl. 1, scale 1:200,000;
1972, Nevada Bur. Mines Bull. (in press).
Nos. 1, 2.